COLLECTING
AND PRESERVING
PLANTS AND ANIMALS

COLLECTING AND PRESERVING PLANTS AND ANIMALS

JENS W. KNUDSEN
Pacific Lutheran University

Harper & Row, Publishers
New York, Evanston, San Francisco, London

Portions of this book originally appeared in *Biological Techniques*. Copyright © 1966 by Jens W. Knudsen.

COLLECTING AND PRESERVING PLANTS AND ANIMALS
Copyright © 1972 by Jens W. Knudsen

Standard Book Number: 06-043744-8

Library of Congress Catalog Card Number: 72-87879

Contents

Preface

This book has been written for courses in biological techniques, field courses in botany or zoology, and for students and others who collect and work with living or prepared specimens of plants and animals for research, teaching, or study projects.

There is much literature on collecting and preserving specimens that has been published over the last three centuries in journals, books, and papers around the world. This paperback, however, provides the more important techniques required for almost all plant and animal preparations. It is a reduced version of the more extensive hardbound textbook, *Biological Techniques: Collecting, Preserving, and Illustrating Plants and Animals,* also published by Harper & Row. For the sake of size, however, the chapter on biological illustrating, diagrams describing the construction of storage cabinets, and the chapters on bacteria, Bryozoa, and the lower chordates have been removed. Thus, the student should consult the parent textbook whenever these techniques are required.

The major reason for writing this book is that of conservation. With this text in each science class, students will be able to preplan their field work and have the proper procedures and solutions for preservation on hand when they return from the field. When specimens are properly preserved, labeled, and stored for present and future use, students will have the satis-faction of knowing that repetitious and wasteful collecting is no longer necessary. As the class collection of carefully preserved specimens becomes adequate for future use, the excitement of locating and identifying plants and animals will continue in the field program. However, a greater pleasure will be derived from releasing or leaving undisturbed nonessential plant and animal specimens. To this end, careful collecting and preservation can foster a greater awareness of conservation.

I am indebted to many people who have made valuable suggestions concerning the updating of this book. Of the many who have contributed, I wish to single out the following: Mrs. Irene Creso, Pacific Lutheran University, who so kindly reviewed the plant chapters; John L. Mohr, University of Southern California, for his enthusiastic review of "cellulose munchers" and other protozoans; Dr. Fred S. Truxal, Chief Curator, Los Angeles County Museum of Natural History, who reviewed the insect chapter; Dr. James H. McLean, also of the Los Angeles County Museum of Natural History, for his many suggestions relative to the mollusks; Dr. David B. Wake, Museum of Vertebrate Zoology, Berkeley, who reviewed the chapter of his specialty, the amphibians; and Dr. Murray L. Johnson, University of Puget Sound, for his comments on the mammals.

I would greatly welcome any suggestions or criticisms concerning both the function of techniques under various field conditions and the overall usefulness of this text. I would also appreciate receiving new or old tech-niques not included herein that the reader feels will give good results, espe-cially in areas of narcotizing and preserving. When corresponding, it would be most useful to have literature citations if these are known.

Tacoma, Washington　　　　　　　　　　　　　　　　Jens W. Knudsen

COLLECTING
AND PRESERVING
PLANTS AND ANIMALS

General Collecting Techniques

Many aspects of sound collecting procedures and some basic marine and fresh-water techniques are applicable to all of the following chapters in this text. Before going afield, a student should know fully what field information, notes, measurements, and techniques will be required. Many fine specimens are lost because needed equipment and supplies were not prepared ahead of time, and the subsequent delay for such preparation after specimens were in hand has permitted spoilage in one form or another. Again, one must always know, prior to collecting, those plant or animal characteristics needed for identification or study so that plants may be arranged and animals can be relaxed or expanded or mounted in such a way as to "preserve" their specific characteristics. Misshapen, distorted, or contracted specimens are usually of little value. Finally, if a collector fails to take full field and color notes, he runs the risk of producing well-preserved specimens that cannot be used by experts. The details of collecting are always so vivid that note-taking is delayed or forgotten. Each collector must, therefore, devise a sound method of taking those field notes essential not only for his use, but for the use of experts should his specimens prove to be rare and valuable. It is for this reason that Chapter 1 considers common collecting and procedural techniques, including those required for the transport of live or preserved organisms, to help ensure against wasted effort and specimens. The parent textbook, *Biological Techniques* (Knudsen, 1966), further discusses common collecting sites such as marine, fresh-water, and terrestrial habitats in order to give additional insight as to what may be expected in the field.

MARINE COLLECTING METHODS AND EQUIPMENT

Exposed Intertidal Zones

Equipment.   The kinds of equipment and personal clothing used on a collecting trip will depend on the zone, the local weather conditions, and the type of plant or animal material sought. Hip boots are excellent for intertidal collecting, in that they permit you not only to wade into deep water, but to kneel down in shallow water without getting wet. If hip boots are not worn, one should plan on getting thoroughly wet and wear tennis shoes which will afford the best footing. On rocky substrates one should have his legs protected and may also benefit from wearing cotton gloves. A little sidepack or knapsack is handy to hold such things as field notebooks, pencils, and plastic bags.

Use plastic buckets or large plastic bags for collecting plants or animals. Metallic buckets are not as suitable in that the metal may kill delicate specimens. Small vials with stoppers or small plastic bags and rubber bands are handy for isolating small delicate species. A pocketknife is useful for prying sessile plants and animals from rocks, but is dangerous and should be used with care on slippery substrates. Remember to oil the pocketknife well before and after collecting. A geology pick is handy for turning rocks and chipping off samples. One should keep his collecting equipment to a minimum and have it well organized so that his hands are as free as possible for collecting.

Low-tide periods frequently occur at night, thus necessitating the use of lights. No portable electric light can surpass the brilliance of the pressure gasoline lantern. The lantern will burn brilliantly for hours and is preferred by the writer, even though it may be damaged by splashing or dropping. Miner's headlights (flashlights) are preferred by many collectors for intertidal collecting, in that both hands are left free. However, flashlights that appear brilliant at home will seem very dim in the intertidal zone, because light is so quickly absorbed by plant material. A supply of batteries is needed: eight to twelve flashlight batteries are required on a single evening's collecting. A spare "emergency" flashlight should always be carried at night when collecting alone, for if you break your light and become stranded in the intertidal zone you will be in extreme danger. The writer, having been stranded on both rocky and sandy beaches at night in heavy fogs without a second light, finally adopted the practice of placing a gasoline lantern on shore above the high-tide mark as a guide to work toward in case of emergency.

Collecting Procedure.   Consult tide tables to select the time for collecting. When collecting on rocky substrates, walk slowly and carefully and crouch low in order to avoid slipping and falling. Attempt to walk on rough rocks or barnacles and avoid smooth rock surfaces, as these may be covered with slippery species of algae. When collecting around breaking waves, learn to count the waves so as to know when large waves will approach and to hear these waves before they reach you. Always collect with a companion or two for safety's sake. When working out on rocky reefs away from the high-tide line, be very conscious of the amount of time that you have been collecting and do not get trapped by the incoming tide. In the Northwest many beach collectors have been trapped against rocky cliffs or out on rocky reefs and have been drowned by the incoming tide.

While collecting in rocky areas observe plants and animals on all rock surfaces and collect only those which are needed. Be careful to turn rocks back to their original position for conservation's sake. Some species of non-mucus-secreting seaweed and a little sea water should be placed in the bucket for receiving animal specimens. Filling the bucket with water is unnecessary, burdensome, and often damaging to the specimen. From time to time flood the bucket with water and then drain the water off again. Be aware of the fact that some species of animals will die and disintegrate very rapidly (especially sea cucumbers and the like), or they may expel sperm and eggs (starfishes), or in one way or another pollute the water and kill other animal specimens in the same container. Some species of plants and animals secrete slime or mucus that may damage other delicate forms. The use of small plastic bags to isolate these species may overcome this problem.

For collecting on sand, mud, or gravel substrates similar techniques are used, although one frequently has to dig and screen large quantities of the substrate in order to locate animal specimens. The need for care not to overcollect and to leave the habitat as undisturbed as possible cannot be overemphasized.

Subintertidal Collecting

Depending on the organisms sought, skin diving works well when aided by hand nets, pry bars, or picks. Plastic bags tucked into swim trunks are used to store specimens; add air, tie, and they will ascend to a waiting skiff. A self-closing net bag is excellent to carry specimens or containers. In collecting from a boat in intermediate depths, various devices such as the dredge, orange-peel bucket, or dippers may be used as described below. Two excellent accounts of bottom samplers by Hopkins (1964) and Thorson (1957) should be read by any serious collector with moderate boat facility.

Dredges and Dredging.   A dredge consists essentially of a strong net attached to a heavy frame which is pulled along the substrate in order to obtain plants and animals. The size of the dredge is determined mostly on the basis of how it will be pulled across the bottom, rather than on the organisms to be obtained. In both fresh-water and marine collecting hand dredging with small gear is very effective; large dredges measuring 8 to 10 feet across may be used only by large fishing vessels. It must be remembered that dredges frequently become caught on rocks and other obstructions and must be somewhat "snag-proof"—or else, expendable. In addition, the net may become torn on rocks or other obstructions and should be protected by a stout outer net made of screen wire, rope, or chain, depending on the substrate and size of the dredge.

In Fig. 1–1, A, B, and C show a small marine dredge with a mouth opening of 8 by 24 inches. The hoop is made of ¼- by 3-inch bar metal but may be made of ¾-inch pipe or any other material. Holes are drilled in the bar metal so that the net can be wired in place. The net may be made of hardware cloth or fish netting. A tow rope is attached to the dredge by means of two arms extending ahead of the hoop and constructed of ⅜-inch rod iron. The distal end of the arm is bent into a ring; the proximal end is triangular and passes through a piece of ½-inch pipe welded to the frame of the dredge. The pipe (Fig. 1–1A) serves as a hinge for the arm. Note that the tow rope does not attach to both arms, but rather to one arm alone (Fig. 1–1B). The second arm is tied to the first by means of medium-weight twine which will break and free the dredge should it become hung up on the bottom, as illustrated in Fig 1–1C. This dredge may be modified by using a rope sling in place of the arms. However, a twine safety device is still important.

There are several ways in which a collector with little equipment may use the hand dredge. The first method is quite simple. Attach one end of a ½-inch rope, 500 feet long, to some object on shore and row out until all of the rope is payed out. Attach the dredge to the rope and drop this to the

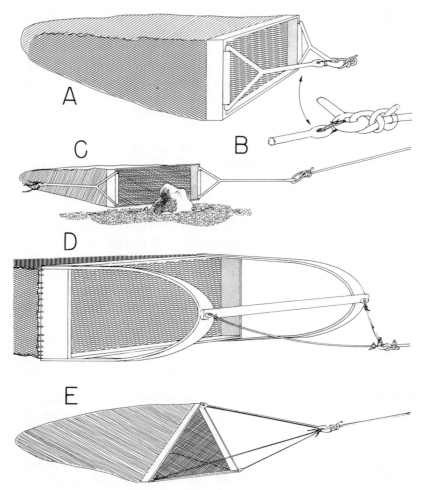

FIG. 1–1. Dredges. A. Biological dredge. B. The detail of securing the two dredge arms by means of string. C. The dredge releases itself from the substratum. D. The frame of a large dredge with a skid. E. A lightweight dredge.

bottom. Return to shore, pull the dredge in hand over hand, and examine its contents on the beach.

Dredging can be done in deeper water without returning to the beach, by pulling the dredge from a skiff equipped with an 18-horsepower outboard motor. However, the dredge must be pulled in by means of a winch or by hand. When dredging in open water, one must be prepared to stop quickly if the dredge becomes hung up on the bottom. Some collectors attach a large float to the dredge rope for emergencies. This is thrown overboard when the dredge becomes hung up and thus lessens the danger of the dredge rope's breaking.

The best dredging method, however, is to attach a large pulley to some object on shore before the dredging operation begins. When the dredge rope is pulled in to shore with the outboard motorboat, a companion on shore quickly passes the end of the rope through the pulley and gives it back to the operator of the boat. The boat then pulls the rope seaward until the signal is given that the dredge has come up on the beach.

Figure 1–1D shows the framework of a large dredge which is provided with a sled and a chain attachment for the tow cable. This kind of dredge will readily foul on rocks or other obstructions, but is very useful on sandy or muddy bottoms.

Figure 1–1E shows a lightweight dredge designed to be used in fresh-water ponds or lakes or in marine collecting. The frame is made of ¼- by 3-inch bar metal built in a triangular pattern. The size of the triangle may be 18 inches or larger on each side. A stout ring is welded to each corner of the triangle. A single 3-foot metal arm made of ¼-inch rod and provided with a ring at either end is attached to one corner of the dredge. Heavy twine or light rope is then attached to the other two corners and tied into the ring as illustrated. Should this dredge become fouled on the bottom the string will break and will usually release the dredge. Quarter-inch minnow netting, some other netting, or hardware cloth may be used for the bag.

Bottom Samplers.   Thousands of species of minute invertebrates living on or in the surface muds can be obtained only by collecting and processing bottom samples. When a population census is being made any sampler that will bring up part of the substrate will be satisfactory, whereas more elaborate gear is needed for quantitative work. Among the few pieces of satisfactory equipment for quantitative work are the orange-peel bucket and the Petersen grab. The orange-peel bucket (Fig. 1–2, A and B) is a modified tool used in construction work. It has four jaws that plunge into the substrate, picking up a particular quantity of mud or sand. The area covered by the bucket is known and the volume of mud can easily be determined; hence the number and kind of organisms per unit of substrate can be determined. Dwarf orange-peel buckets are available commercially (Hytech Corp., 6803 West Boulevard, Inglewood 3, California, and others). These weigh 35 pounds, have a capacity of 100 cubic inches, cover an area of 11½ inches in diameter, and are very satisfactory when used from a skiff.

Figure 1–2C shows a scoop sampler designed for the U.S. Public Health Service which is made from any convenient container with a U-shaped metal

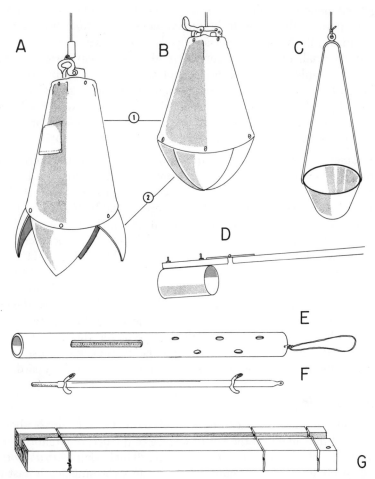

FIG. 1–2. Other collecting equipment. A–B. The orange-peel bucket opened and closed. (1) The canvas cover which prevents the loss of the sample, (2) the jaws. C. A dipper for bottom sampling. D. A tin-can bottom sampler. E. A tubular thermometer holder. F. The thermometer supported by two pieces of rubber tubing. G. A wooden thermometer holder.

arm and a ring for attachment. This is pulled along the bottom until a sample is obtained. One of the drawbacks of this device is that the surface area covered is unknown and some of the organisms may become lost as a sample is pulled to the surface.

A shallow-water sampler designed by the writer is shown in Fig. 1–2D. This consists of a tin can bolted to a short stick. This, in turn, is hinged to a long stout handle which can be maneuvered from shore or from a skiff. The hinge permits the scoop to be pulled along the bottom from any angle and thus makes the sampler a versatile collecting device.

Preserve a small quantity of surface mud in 5-percent formalin and another small sample in 75-percent alcohol so that microscopic plants and animals can be removed in the laboratory. These samples should be labeled as to locality and depth of water in which they were taken. Wash the remainder of the sample on a series of screens so that the plants and animals present can be picked out and preserved in appropriate solutions.

Thermometers.   In any kind of aquatic or marine collecting thermometers will be useful when complete data are being taken. Conventional laboratory thermometers reading in $\frac{2}{10}$ of $1°$ C. are quite suitable. Small stick thermometers already enclosed in a metal case are available. Figure 1–2E shows a protective case made out of $\frac{1}{2}$-inch metal or plastic tubing which has a slot cut in it to permit the reading of the thermometer. The thermometer is held by two short pieces of rubber surgical tubing (Fig. 1–2F) and is tied in place by means of a string. Another kind of thermometer holder can be made by ripping a groove in a small board which is slightly longer than the thermometer (Fig. 1–2G). Make three shallow saw cuts across the board. Tie the thermometer firmly in place, passing the string through two of these cross cuts. Then pass a string through the eye of the thermometer and tie it in the third cut. The groove receiving the thermometer must be deep enough that the thermometer will lie below the surface of the board.

Pelagic Collecting

The pelagic zone of the ocean consists of all of the water mass above the substrate and thus includes the entire surface of the ocean. Organisms living there may be planktonic—that is, organisms which simply drift with the water—or nectonic, that is, organisms which can swim sufficiently fast to be independent of ocean currents. In the latter category fishes and squid are the chief forms while thousands of species of both adult and larval plants and animals are found in the plankton. Large tow nets or night lights are used for collecting necton, and plankton nets are used for the smaller forms.

Plankton Nets.   Plankton nets are usually $9\frac{1}{2}$ inches in diameter and 35 inches long, tapering to a point. The meshes vary from very coarse to very fine, ranging from 20, 40, 75, 125, 175, to 200 meshes per inch. Silk is the standard material used in making plankton nets, and muslin is used to reinforce the ring. The tapered tip of the net is either tied shut or tied around a small vial which collects the plankton. Materials for making plankton nets are available, but handmade nets are less suitable than those purchased from biological supply houses. Because of the high cost of such nets (fifteen to fifty dollars) they should be carefully washed in fresh water and air dried to prevent rotting or other damage.

Plankton nets are either towed through the water on the end of a heavy cord or thrown out from a skiff and then retrieved. They can be made to "fish" at various depths by adding weights just ahead of the net and by carefully controlling the length and angle of the line. Another method of using the weighted plankton net is to drop it straight into the water and retrieve it repeatedly until a sample is obtained.

Night Lighting.   One rather exciting form of collecting is "night lighting," which consists of hanging a light near the surface of the water from the side of a ship or the edge of a pier. This technique works best where no other lights will distract the marine organisms. After a brief period of time plankton will slowly work its way up toward the light and can be removed with a plankton net or dip net. Small and large fishes, squid, and even swimming crabs will appear under the night light. The composition of the species will change every few hours during the night so that one can profitably collect all night long.

Bulk Field Preservation

Dredges, deep nets, plankton nets, and intertidal collectors all gather large quantities of material indiscriminately. Not all of this material can be preserved in the same way, nor is it desirable to attempt to preserve all of the specimens. An outline of some of the choices that must be made in the field with regard to the treatment of specimens is as follows:

A. Specimens to be kept alive for transportation to the laboratory for subsequent culturing or preservation must be isolated.
B. Plants and animals to be preserved in the field must be sorted as follows:
   1. Animals needing narcotization (those which contract or autotomize) must be sorted for the several probable techniques to be used and subsequently preserved by the appropriate method.
   2. Organisms requiring direct preservation must be sorted as follows:
      a. Specimens to be preserved in alcohol.
      b. Specimens to be temporarily or permanently preserved in formaldehyde solution, either normal or neutralized.
      c. Specimens to receive special fixation for tissue or anatomical studies.

Each group of plants or animals may require some special technique depending on its nature and the intended use of the specimen. Often huge quantities of specimens are preserved in the field en masse with 5-percent to 10-percent formalin (preferably neutralized). Formalin will affect specimens with calcium carbonate structures, but is usually quite suitable for temporary preservation. In the laboratory these specimens are washed, sorted, or soaked out, and reassigned to permanent preservative solutions according to their particular need.

Treatment of Plankton

Because the content of the plankton net will be a conglomerate of various species, it is advisable to preserve small portions in 5-percent formalin, in hot (50–60° C.) FAA, and in alcohol. Formalin and alcohol will cause many of the forms to contract, whereas hot FAA poured over the concentrated plankton will kill numerous specimens in an expanded condition. Because the alcohol may cause precipitation of salts from the sea water, a change of preserving solution should be made soon after the initial preservation. Small vials with stoppers that will not trap and hold the plankton (as will cotton or porous corks) should be used to hold the plankton. These, in turn, are placed in screw-top jars after the proper collecting data have been enclosed.

Slide Preparation.   In the laboratory plankton may be stored in vials or portions may be prepared on slides. When the latter technique is used masses of plankton may be stained and generally prepared on slides or individual specimens may be selected from the stained material and isolated on slides. Several stains are generally used: borax carmine, iron-hematoxylin, fast green, and the like. Place plankton to be stained in a small vial and wash several times with oxygen-free water (previously boiled and cooled). Pour off most of the water and add stain. After about an hour carefully pour off stain, add fresh oxygen-free water, and, after the plankton has had time to settle, pour off the water and replace with 50-percent ethyl alcohol. Dehydrate and infuse with xylene, and mount in balsam or Permount as directed in Appendix B, or omit dehydration and mount directly in Turtox CMC-10. Plankton may also be mounted directly from water or alcohol in CMC-S which stains and mounts (see Appendix B).

Transporting Live Plants and Animals

Some general problems encountered in transportation are (1) maintaining proper temperatures, (2) reducing the metabolic rate, (3) supplying adequate amounts of oxygen, (4) preventing pollution of sea water or fresh water, (5) preventing overcrowding and (6) allowing for the general incompatibility of different kinds of organisms. The time of day, the air temperature, the distance the organisms must be transported, and the number of organisms, all help to determine the methods of transportation.

Containers.   Avoid using metal containers, unless they have a porcelain lining, as they will give off toxic substances. Plastic or glass is best, as neither will contaminate the specimens. (Some notable exceptions for glass containers will be mentioned in the discussion of fresh-water algae.) When delicate animals are transported one should include some nonmucus-secreting algae which will provide hiding places and prevent damage from motion. Small arthropods such as crabs are prone to fight with one another unless they are provided with hiding places. Do not overcrowd the container with either plants or animals. Most marine invertebrates which normally survive exposure at low tide are best transported in wet seaweed with almost no water.

Aeration en Route.   If specimens are to be transported in water, place only a few inches of water in each container to maintain a high ratio of surface area to volume; this permits a greater degree of diffusion of oxygen. Car motion will often create small waves in the container, facilitating oxygen diffusion. The presence of some plants will also provide oxygen for short periods of time, but cannot be relied upon for long shipping periods. If water temperatures become higher than those of the habitat additional aeration is essential. Mechanical agitation of the water or pumping of air through the water by means of a hand syringe or tire pump may be essential. The best technique for aeration is as follows: Take a superinflated inner tube, a length of rubber tubing, and an aquarium stone along on the collecting trip. When you are ready for aeration, partially unscrew the valve to release air, place the rubber tube over the valve stem, and stick the aquarium stone in the opposite end of the rubber tube. Place the aquarium stone in the container and, if necessary, readjust the inner-tube valve until the proper flow of air is achieved. Stop at as many gas stations as necessary to reinflate the inner tube en route. Battery-driven aquarium pumps or even cylinders of compressed oxygen may be substituted if they are available.

Temperature Control

Styrofoam ice chests are excellent for maintaining low temperatures while transporting specimens isolated in jars or plastic containers. Cooling may be maintained by adding small quantities of ice, packed so that ice does not touch the containers themselves. The ice chest may serve as a water bath when marine specimens are transported in jars. Water is periodically replaced by cooler water, unless ice is used. Caution is required: use ice sparingly; do not allow the bath water to enter specimen containers.

FRESH-WATER COLLECTING
METHODS AND EQUIPMENT

General Equipment

Collecting methods and equipment must be geared to the organisms sought. The aquatic dip net proves invaluable for collecting swimming organisms and for bringing debris to the shore. The dredge net is designed to collect large quantities of debris from shallow water. Such debris is sorted on shore and will yield numerous species of plants and animals. Special dippers, kitchen strainers, or aquarium nets are also useful for gathering isolated specimens. In running water the stream net and dip net are useful for catching specimens that have been dislodged from their hiding places. The construction and use of these nets is described in Chapter 12. Occasionally, a large pipette is useful for getting bottom samples. This apparatus is made by placing a 50-cubic-centimeter rubber bulb on the end of a long length of plastic tubing. The plankton net, biological dredge, and scoop samplers described above are considered among the most important fresh-water collecting devices. See Welch (1948) for a thorough discussion of the techniques used in fresh-water ecology and measurement.

General Collecting Methods

In collecting within the permanent aquatic habitat utilize the plankton net for sampling the pelagic plants and animals. These organisms are highly cyclic; they will change from week to week throughout the season and require a continuing program of sampling. The number of organisms near the surface may also vary daily with changing light conditions or with minute but rhythmic temperature fluctuations.

Aquatic vegetation harbors a tremendous number of invertebrate species. These populations may differ greatly between those associated with plants growing on the surface and those harbored by plants growing in deep water. In shallow water carefully cut vegetation free from the substrate and float it into the dip net without disturbing the organisms. Next, hunt through the vegetation with a field dissecting microscope or hand lens for attached forms. Place similar vegetation in a clean plastic bucket of water and add a small quantity of formalin. This will drive the motile forms from the vegetation and ultimately kill them, permitting them to settle to the bottom. Carefully remove the vegetation and filter the fluid to obtain the specimens.

When working along the bottom substrate, carefully remove rocks and examine both the upper and under surfaces. Large sticks and logs should be checked, the bark removed, and, if necessary, the log broken open to expose the hordes of specimens living there. Small pieces of gravel should be examined underwater with the hand lens. Large quantities of organic bottom debris should be brought in with a dredge net and carefully picked through on the shore, or placed in a screen and washed, or examined in small quantities with the dissecting microscope or hand lens.

When collecting from stream bottoms, place a stout dip net against the bottom and carefully lift out stones so that any animals that are dislodged will float down into the net. Each time the stream widens or narrows, deepens or shallows, has a change in slope or in the nature of the substrate, there will be a change in the animal and plant population. All macroscopic and microscopic crevices should therefore be examined.

A strikingly different population of plants and animals may be found in temporary ponds or puddles. If the pond is dry, collect samples of the uppermost bottom mud with a hand trowel or pocketknife and put them into filtered pond water. After several days or weeks at room temperature many plants and invertebrate organisms which were present in resistant stages will appear. Otherwise, temporary water is sampled with the same techniques as that of permanent lakes or ponds.

SHIPPING
PRESERVED SPECIMENS

Dry Specimens

Dried specimens such as sponges, echinoderms, coral, algae, and the like, should be carefully supported by tissue paper in individual cardboard boxes with the data included. These boxes, in turn, are placed in a large, strong cardboard box and completely surrounded by crumpled newspaper or other soft packing. Paradichlorobenzene crystals or naphthalene flakes may well be included as a fumigant, especially if the material must go through customs. The purpose of packing dry material is to prevent shaking and jarring (tissue paper support around the inner box) and crushing (the outer box). See Chapter 12 for methods of shipping dried insects.

Specimens Preserved in Liquid

Specimens preserved in liquid present problems of weight and leakage. Material such as formalin may not be shipped through the regular mail and will have to be sent by freight. The size of the organisms must be considered first, and will be handled in one of three ways: (1) Very small specimens must be kept in vials (preferably plastic) with small quantities of preservative, data, and a cotton stopper or screw cap. An alternative is to place the specimens, data and preservative in a small plastic freezer bag. (2) Specimens of small and intermediate size may be grouped according to collecting locality and date, wrapped in cheesecloth (along with the field data), and tied with a piece of string. (3) Large specimens, such as amphibians or fishes, may be individually tagged with field data.

Once the specimens are processed with their data they should be placed in a stout plastic bag, along with some additional paper toweling and a small quantity of the preservative. Carefully tie the top of the plastic bag after expelling most of the air and place the bag inside a screw-cap or snap-cap tin can or metal drum. If necessary, add padding so that the specimens will not bounce around inside the drum. An alternative is to put the bag into a second and third plastic bag, each sealed in turn, and to ship this in a pasteboard box padded with crumpled newspaper.

Over the past fifteen years many field collectors in museums have used regular home-canning devices for the packing and shipping of specimens in tin cans. Preserved specimens are prepared as described above, placed in cans, padded, and sealed. Taylor (1950) thoroughly reviews the techniques required and mentions that many kinds of plants and animals including coral, echinoderms, algae-encrusted rocks, fungi, soil samples, and the like, can easily be shipped in this manner.

FIELD NOTES

Specimens soon become worthless unless they are accompanied by adequate data. Each collector should therefore keep a notebook in which the field data are arranged chronologically, in addition to putting slips of data in the specimen containers. You may choose between a bound, field notebook of good quality, high-rag-content paper or the looseleaf variety. Collectors using the looseleaf notebook transfer their field notes into a master notebook which always remains in the laboratory, and thus avoid the risk of losing past data. There are many suitable field notebooks on the market, but one of the most satisfactory for work in the Pacific Northwest is the "Rite in the Rain" #311 (Darling Corp., Tacoma, Washington), which is entirely waterproof. The paper offered by this company can be used underwater or in pouring rain with equal facility, will not wrinkle badly, and will always dry out and look very respectable. Notes should always be taken with a medium-grade pencil or with waterproof ink—not with ink that will run if subjected to water.

Three types of data should always be taken for complete notes. First, the locality according to state, county, nearest township, and approximate location in miles distance from the nearest township should be given, with the idea in mind that you or some other collector may wish to locate the spot with little difficulty. Do not use local names that are not established on land maps, as they will soon become valueless. By like token, do not simply give the name of a bay or town or lake; if you do your specimens may eventually have to be discarded for lack of adequate data. For example, there are seven "Clear Lakes" in the state of Washington, three lakes in one county alone bearing this title. Thus "Clear Lake" specimens are valueless unless additional data are given. Second, describe the habitat in which you collect in detail sufficient to help a new collector locate the locality. Kinds of vegetation, whether you are dealing with a meadow or forest or river, and so on, should be noted. Third, data concerning the kinds of specimens and the conditions under which they were collected, including weather conditions, hiding places, kinds of activity, and so on, should be jotted down.

COLOR NOTES

Occasionally it is necessary to make color notes in the field of some plant or animal whose color will be destroyed or altered before it can be examined in the laboratory. Robert Ridgway (1912) prepared an excellent but elaborate color standard as did Maerz and Paul (1950). Both of these books have colors arranged in small squares across and down the page, each keyed alphabetically and numerically so that the worker may match the color square with the actual specimen and code the color in the field notebook. These color standards are available only through libraries. Neither, however, is ideal for actual field collecting. A second popular alternative to color standards would be the use of color photographs. However, unless some known color standard, such as the color cards for house paints or water colors, is included in the photograph, the accuracy of the photo cannot be guaranteed. With a color standard the photograph can be compared with the standard in the laboratory and a judgment made of the degree of difference existing between the photograph and the standard.

By far the simplest and most direct method of making field color notations is to use an inexpensive box of watercolor paints. This comes equipped with a small brush and mixing pan and is small enough to pack in a field kit. First make a crude sketch of the specimen, showing the areas where particular colors are located. If, for example, you have a crab such as *Grapsus grapsus,* which is orange-red on the back and light sky blue on the undersurface, mix a small quantity of red and yellow and paint a small area of this on the margin of the paper. When you hold the paper against the specimen itself you will quickly see whether the paint blotch is too red or too orange, whether you should add yellow or blue to make it lighter or darker, and so on. By experimentation and frequent comparision of colors you will eventually get one color blotch that actually matches the specimen. Encircle this color blotch and put an arrow to the place on the specimen that it matches most exactly. Use the same procedures for the light blues or other colors that may be present. Do not, unless you wish, attempt to paint the drawing you have made. In the laboratory the color blotches may be analyzed and more carefully described for publication.

REFERENCES

Allee, W. C., *et al.,* 1949, *Principles of Animal Ecology,* Saunders, Philadelphia.

Buchsbaum, Ralph, 1948, *Animals Without Backbones,* Univ. of Chicago Press, Chicago.

Clarke, G. L., 1965, *Elements of Ecology,* Wiley, New York.

De Latil, P., 1955, *The Underwater Naturalist,* Houghton Mifflin, Boston.

Ekman, S., 1953, *Zoogeography of the Sea,* Sidgwick & Jackson, London.

Hausman, L., 1950, *Beginner's Guide to Fresh Water Life,* Putnam, New York.

Hopkins, T. L., 1964, A Survey of Marine Bottom Samplers, *Progr. Oceanogr.* 2:213–256.

Knudsen, J. W., 1966, *Biological Techniques: Collecting, Preserving, and Illustraing Plants and Animals,* Harper & Row, New York.

MacGinitie, G., and N. MacGinitie, 1968, *Natural History of Marine Animals,* McGraw-Hill, New York.

Maerz, A., and M. R. Paul, 1950, *Dictionary of Color,* McGraw-Hill, New York.

Miner, R. W., 1950, *Fieldbook of Seashore Life,* Putnam, New York.

Morgan, A., 1930, *Fieldbook of Ponds and Streams,* Putnam, New York.

Needham, James G., *et al.,* 1937, *Culture Methods for Invertebrate Animals,* Dover, New York.

Needham, J., and P. Needham, 1962, *A Guide to the Study of Fresh Water Biology*, Holden-Day, San Francisco.

Ricketts, E. F., and J. Calvin, 1968, *Between Pacific Tides*, Stanford Univ. Press, Stanford, Calif.

Ridgway, Robert, 1912, *Color Standards and Color Nomenclature*, published by the author, Washington, D.C.

Smith, Ralph, *et al.*, 1954, *Intertidal Invertebrates of the Central California Coast*, Univ. of California, Berkeley, Calif.

Sverdrup, H. W., *et al.*, 1942, *The Oceans*, Prentice-Hall, Englewood Cliffs, N.J.

Taylor, William R., 1950, Field Preservation and Shipping of Biological Specimens, *Turtox News* 28(2):42–43.

Thorson, G., 1957, Sampling the Benthos, J. W. Hedgpeth, Ed., Treatise on Marine Ecology and Paleoecology, *Geol. Soc. Amer. Mem.* 67(1):61–68.

Ward, H. B., and G. W. Whipple, 1945, *Fresh-Water Biology*, Wiley, N.Y.

Welch, P. S., 1948, *Limnological Methods*, McGraw-Hill, Blakiston Division, N.Y.

The Algae

That group of plants collectively known as "algae" consists of seven distinct phyla found in the subkingdom Thallophyta. Thallophyta (a group which also includes the bacteria and fungi) have simple plant bodies with little cellular definition; they lack roots, stems, and leaves which are comparable in cellular structure to the higher plants. This simple plant body is referred to as the thallus. Unlike bacteria and fungi, the algae carry on photosynthesis and possess photosensitive pigments of various colors. The seven phyla of algae are thus popularly named after their general coloration and may be referred to as the green algae, yellow-green algae, brown algae, red algae, and so on.

Many references are cited at the end of this chapter for information on algal identification. Culturing algae is often important before study and research may continue; see Fogg (1966) and Pringsheim (1967). See also *Biological Techniques* (Knudsen, 1966) for culturing techniques and methods of working with diatoms. Refer to "Subintertidal Collecting" in Chapter 1 of this book for collecting methods and to Taylor (1967) for an excellent discussion on prolonged field collecting.

MARINE ALGAE

Algal Habitats

In surveying a long expanse of coastline along any of the continents one may be impressed with the seemingly repetitious appearance of headlands, rocky beaches, sandy beaches, bays, and estuaries. To the trained eye, however, tremendous ecological changes occur between the latitudes, especially in light and temperature. These two factors, along with the presence or absence of nutrients, are most important in controlling plant growth and distribution. Some species of plants or animals may seem to occur almost everywhere along a coastline, but generally there is a gradual progression of species, some dropping out and new ones being added as one moves toward the north or south.

Morphology of the Algae

Generally speaking, the larger species of algae grow in the cooler temperate water, whereas the stature of most species seems to diminish as one approaches the warmer tropical water. There is great variety in shape and form, ranging from the extremes of erect, free-growing plants to those which encrust on the surfaces of rocks. All of the erect forms have a holdfast for attachment, and may include a stipe (a stem-like structure), a float, and one or more blades (leaf-like portions) as shown in Fig. 2–1A. The holdfast may be a simple, nipple-like growth attached to the substrate or a complex root-like mass. The erect portion may range in morphology from a filamentous, hair-like mass to a large flat blade, or even a sac-like mass, as seen in Fig. 2–1, B, C, and D. Many species of erect algae deposit lime in the thallus, and thus consist of hard, jointed segments (Fig. 2–1E). Quite a few species of both fleshy and coralline (hard, stony) algae grow directly on rocks, in a prostrate, encrusting form and their removal may require the use of a geology pick and chisel (Fig. 2–1E).

Collecting Marine Algae

Equipment.   Among the few pieces of equipment required for collecting marine algae are a plastic bucket or a very large, stout plastic bag for transporting specimens, some small plastic bags, rubber bands, a few vials (preferably plastic) for isolating minute specimens, a pocketknife, and a field notebook. A geology or mason's hammer is useful in collecting and turning rocks. Occasionally, a cold chisel is essential for chipping off bits of rock

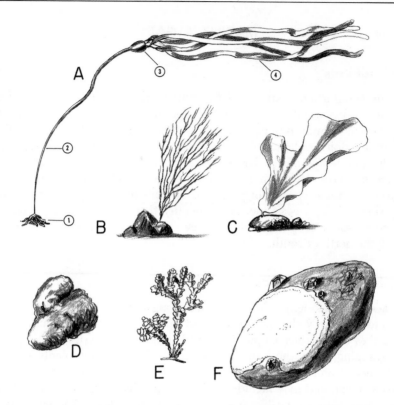

FIG. 2–1. Algal morphology. A. Large kelp. (1) Holdfast, (2) stipe, (3) float, (4) frond. B. Filamentous alga. C. *Ulva,* a sheet form. D. *Culpomenia,* sac-like. E–F. Coralline algae, erect and encrusting.

containing encrusting seaweeds. Finally, a protected thermometer is desirable when ecological interests are involved.

Collecting Techniques.   Plant specimens of a given species will vary greatly in appearance, depending on the season of the year, the age of a given plant, and the suitability of the habitat in which the plant lives. Whenever possible, collect a series of specimens of each species, attempting to get mature, well-formed plants. One or two specimens will suffice for larger species. Look carefully at the growth of algae surrounding you, so as to include "look-alike species."

In the intertidal and accessible subtidal areas, carefully remove the specimens by slipping a pocketknife underneath the holdfast. Put delicate forms in vials or small plastic bags with a small quantity of water. Carefully select specimens of encrusting algae which grow upon small single rocks or from parts of rocks that can be readily dislodged with a chisel and hammer. Many epiphytes (plants that grow upon other plants) will be found among the algae and should be included. Plants are usually transported to the laboratory for preservation, but occasionally it is essential to temporarily dry or preserve specimens when finished processing is impossible.

When collecting, be sure to collect only enough specimens to satisfy your requirements, to return rocks to their normal position if they are overturned, and, in general, to disturb the habitat as little as possible. There is a great danger not only in overcollecting plant specimens, but in disturbing the habitat for invertebrate animals.

Following severe marine storms great quantities of seaweeds are loosened from their substrata and transported in toward shore. Windrows of weed will occur along the beaches and the tide pools will be filled with loosened specimens. Collecting after storms is very rewarding in that you can get those hard-to-obtain specimens from rocky coasts. Collect on sandy beaches as well as on rocky coasts, for the flora will vary offshore from place to place.

Color and Field Notes.   The colors of marine algae will quickly change in the process of dying, drying out, or on exposure to light. Some species of the red and brown algae will turn sufficiently green, if they die slowly in stagnant water, so that even the phylum becomes doubtful. Some preliminary sorting by color into the respective phyla may be essential. The final processing and preservation must take place as quickly as possible to prevent color alteration. When specimens are to be stored in liquid, put them in a dark place to prevent color alteration. Color notes and complete field notes should be taken if the specimens are to be of the greatest scientific value (see Chapter 1).

Preservation

Liquid Preservation.   Liquid preservation has a number of distinct uses; a few examples of uses follow. (1) It is desirable to preserve a small number of specimens of each species in liquid, at least until they are identified and especially if the species is quite small. (2) Specimens may be briefly held in 5-percent formalin solution until they can be properly mounted in the laboratory. (3) Large fleshy floats of some of the species of kelp may be suitably preserved in formalin for permanent specimens. (4) Specimens may be preserved for classroom dissection, then soaked for twenty-four hours in fresh water prior to use.

When preserving huge quantities of specimens in small containers, increase the percentage of formalin. Use an FAA for preserving specimens intended for slide-making and histological studies, in the ratio of 93 parts of formaldehyde to 3 parts of glacial acetic acid. One of the drawbacks of liquid preservation is that some species become soft and many will lose their natural coloration, especially if they are exposed to light. Therefore, tightly cork or seal vials containing specimens, along with their field data, and place these vials in a large gallon bottle which is in turn sealed. This double-sealing technique will prevent rapid evaporation and permit simple storage of many vials in a dark place.

Herbarium Mounts.   The materials used for herbarium mounts—including papers, methods of attaching specimens to the sheets, methods of constructing and using plant presses, and so on—are discussed in Chapter 5 in sufficient detail.

Smaller specimens of marine algae should be mounted directly. Arrange the more rigid specimens on herbarium sheets in such a manner as to show the important taxonomic features. Most small species of algae will "glue" themselves to the herbarium sheets during the process of drying. Cover the exposed surface of the algae with cloth, such as sheeting, muslin, or cheese-cloth, to prevent the specimens from sticking to the dryers. These covers are removed when the specimens are dry.

Very flimsy specimens must be floated on to the herbarium sheet in the following manner: Float a specimen in a large pan of sea water and carefully slide a herbarium sheet under it. When the plant is arranged in a suitable fashion slowly lift the sheet from the water, with the plant intact. Drain off the excess water by tilting the sheet, cover the plant with cloth, and press lightly, until dry. Herbarium sheets are frequently cut into suitable sizes, such as 4 by 6 inches, onto which small specimens are mounted. These are later kept in a card file or, preferably, mounted on a herbarium sheet.

Rolled Specimens.   Some of the giant species of kelp, which have fronds measuring up to a foot or more in width and measuring from 9 to 90 feet in length, are impossible to mount on herbarium sheets. Small specimens of such species may be selected, and from them representative sections may be mounted in the normal manner. However, intact plants, including the holdfast, may be preserved by either of two methods. Guberlet (1961) calls for soaking specimens in 10-percent carbolic acid and 30-percent each of glycerin, alcohol, and water. When saturated, dry specimens, after which they remain flexible for a long time. Dawson's method (1956) calls for drying giant kelp on newspapers in the shade, then rolling while yet flexible, beginning with the holdfast. Rolls are secured with string, labeled, and wrapped in paper; then left out to finish drying. When dry, specimens prepared by either method are stored in boxes. For reexamination soak out Dawson specimens in sea water to obtain life-like coloration and texture.

Coralline and Other Encrusting Algae

The hard coralline algae and those species encrusting upon rocks present a new problem in preservation. Erect species of coralline algae should be spread out while still fresh, arranged in suitable position without too much overlapping of the branches, and permitted to partially dry. They are then dipped into herbarium glue on a glass plate (Chapter 5) and attached directly to herbarium sheets or cards where they will complete their drying. The alternative method is to store such specimens, directly or on cards, in herbarium envelopes (described in Chapters 4 and 5).

Select encrusting species of algae attached to rocks which are small enough to be stored in cardboard boxes within the herbarium tray. Chips of rock may be glued to herbarium cards and kept in herbarium envelopes or attached directly to the herbarium sheets.

Slide Techniques.   In the identification of algae it is essential to make temporary slides from time to time to demonstrate the cellular structure of various parts of the plant. The arrangement of cells and the thickness of the various tissues are frequently characteristic of families, genera, and species. Mount whole fragments of plants in sea water under a coverslip for study. Make cross sections by holding the plant down with one hand and slicing extremely thin sections with a single-edged razor blade. If the specimen is large and fleshy, and therefore difficult to slice, permit it to dry and then slice the section. When the cross section is returned to a slide in a drop of sea water it will resume its natural proportions and be ready for study. Usually no stain is needed, since fresh or dried materials contain their natural pigments.

Specimens for permanent slide mounts should first be killed in 5-percent formalin or in FAA, as stated above, and then washed in fresh water to remove the fixative. Dawson recommends, in his book (1956) and in his course in Marine Botany, the following procedures: Stain specimens with acid fuchsin or analin blue while they are on the slide, acidify and wash out the excess stain. Blot away the surplus water and add a 60-percent solution of clear corn syrup. Let this stand for 24 hours in a dust-free place and then replenish the evaporated syrup with an 80-percent solution. For thinner specimens, place a coverslip on at this time and, after labeling, store the slide in a flat position. Thicker specimens may be allowed to stand 24 hours longer, then have a slight additional quantity of 80-percent syrup added along with the coverslip. Care must be taken not to dislodge the coverslips or to turn the slide on edge. After a month or two, however, such slides may be considered permanent.

FRESH-WATER ALGAE

Fresh-water algae are usually small and delicate as compared to the marine species. Therefore, living cultures, when it is possible to maintain them, are often preferred to preserved specimens. Algae are usually cultured until the desired state of growth is obtained; they may then be preserved or dried while attached to the normal substrate or, in the case of some of the larger species, attached to herbarium cards. Usually the first two spring months are the best time for collecting pure colonies of algae; the summer months present mixed cultures consisting of two or more species.

Location and Collection of Fresh-Water Algae

Small aquatic specimens are collected directly into bottles or vials, provided with just enough pond water to cover the specimens; containers filled with water will soon stagnate and destroy the samples. Larger specimens, or objects containing colonies of algae, may be placed in plastic bags or rolled into wet (and then dry) newspaper for transportation to the laboratory. Submerged objects in ponds, lakes, and streams provide excellent habitats for algae. Rocks, bits of wood, sticks, and other debris should be removed and examined if they have a green or brown appearance, as this often denotes the presence of algae. These growths can be scraped into vials with a dull pocketknife or fingernail. In the late spring and summer months colonies of floating filamentous algae will appear in great quantities.

Many substrates near ponds and streams, or within humid forests, will harbor flourishing colonies of algae. Greenish soils may be picked up on a piece of tin and transferred into a bottle lid. The bottle is then screwed into the lid and kept in the inverted position while it is transported back to the laboratory. Remove bits of bark containing green algae with a pocketknife. Areas such as water seeps along road banks, bird baths, fish ponds, and the like may also provide sources of less common species of algae.

Preservation of Fresh-Water Algae

Although fresh-water algae are generally preserved in liquid, larger species may be floated on herbarium cards, pressed, and dried according to the procedure described for marine algae. For liquid preservation pour off the collecting water and replace with 3-percent or 5-percent formalin, or FAA. Add labels containing the field data, cap the vials, and store them in a dark place in airtight jars. The more delicate specimens of algae may be prepared on slides, as described above under marine algae.

MUSEUM
AND STORAGE METHODS

Proper storage facilities must be provided in order to conserve not only the collector's time and expense, but the valuable specimens as well. The time has come when repetitious collecting must be avoided in order to conserve and maintain natural plant and animal populations. Specimens preserved in liquid should be tightly sealed, along with internal labels containing field data, and kept in a dark place. Vials should be placed in airtight jars of a convenient size to prevent excessive evaporation. Specimens mounted on herbarium sheets should be carefully labeled and sorted by family, genus, and species. If the collection is sufficiently large, a card file or some other adequate reference should be kept. Standard herbarium cases are readily available at biological supply houses, but can be simply and very inexpensively constructed. These may be made in the home or school workshop for a fraction of the commercial cost, as described fully in *Biological Techniques,* Chapter 5 (Knudsen, 1966).

REFERENCES

Dawson, E. Y., 1956, *How to Know the Seaweeds,* Brown, Dubuque, Iowa.

Faridi, M. A., 1964, Some Simple Techniques for the Production of Zoospores in Fresh-Water Algae, *Turtox News* 42(2):58–59.

Fogg, G. E., 1966, *Algal Cultures and Phytoplankton Ecology,* Univ. of Wisconsin Press, Madison.

Forest, H. S., 1954, *Handbook of Algae: Special Reference to Tennessee and Southeast,* Univ. of Tennessee, Knoxville.

Guberlet, M. L., 1961, *Seaweeds at Ebb Tide,* Univ. of Washington Press, Seattle.

Johansen, Donald A., 1940, *Plant Microtechnique,* McGraw-Hill, New York.

Knudson, J. W., 1966, *Biological Techniques: Collecting, Preserving, and Illustrating Plants and Animals,* Harper & Row, New York.

Nielson, J. E., 1944, Meet the Diatoms, *Turtox News* 22(2):40–43.

Prescott, G. W., 1962, *Algae of the Western Great Lakes Area,* Brown, Dubuque, Iowa.

Prescott, G. W., 1970, *How to Know the Freshwater Algae,* Brown, Dubuque, Iowa.

Pringsheim, E. G., 1967, *Pure Cultures of Algae,* Hafner, Darien, Conn.

Sass, John E., 1958, *Botanical Microtechnique,* Iowa State Univ. Press, Ames, Iowa.

Smith, G. M., 1950, *The Fresh-Water Algae of the United States,* McGraw-Hill, New York.

Smith, G. M., 1955, *Cryptogamic Botany, Vol. I: Algae and Fungi,* McGraw-Hill, New York.

Taylor, W. R., 1957, *Marine Algae of the Northeastern Coast of North America,* Univ. of Michigan Press, Ann Arbor.

Taylor, W. R., 1967, *Marine Algae of the Eastern Tropical and Subtropical Coasts of the Americas,* Univ. of Michigan Press, Ann Arbor.

Tiffany, L., and M. Britton, 1971, *The Algae of Illinois,* Hafner, Darien, Conn.

The Higher Fungi

The fungi comprise a group of primitive plants, usually devoid of chlorophyll, which utilize organic matter for food. The techniques for bacteria (often included with the fungi by many workers) are treated in *Biological Techniques* (Knudsen, 1966) and elsewhere. This book will restrict itself to lichens, mushrooms, and higher fungi.

LICHENS

The lichens are symbiotic plants composed of fungi and algae. They are abundant, found growing directly on soil, rocks, bark, and other substrata. While common in humid or moist areas, especially around forests, they are not uncommon in harsh desert regions. They occur in three classical forms: *crustose,* which totally adhere to the substratum; *foliose* or leaf-like; and *fruticose,* which branch and are pigmented on all surfaces.

Collecting Lichens

Collect and transport specimens in plastic bags. A knife may prove useful for dislodging pieces of bark containing lichens, for digging up soil-dwelling forms, for cutting branches with lichens, and so on. A geology pick and chisel will be essential for chipping off rocks containing crustose lichens. The field notebook should always be carried along for recording of habitat data (see Chapter 5).

Those species of lichens which grow on trees, bushes, mosses, and the like, are removed from the substrate when possible or collected attached to small portions of the substrate. Be careful to get at least some of the rhizoids, the root-like structures which grow on the under side of the lichens and attach them to their substrate. When digging lichens directly from the soil wash them to free the rhizoids from the soil. A thorough washing in the field will greatly simplify the cleaning preparation in the laboratory. Substrata other than soil are often desirable, in small quantities, in order to show the specimen as it was growing in nature.

Those species which grow directly on the surfaces of rocks are more difficult to collect. By carefully looking over the terrain you may find rock specimens small enough to be collected directly. Others may be conveniently sampled with a coal chisel and hammer. Place the chisel in such a way as to fracture off a thin section of rock. Wear glasses or goggles to protect your eyes from rock and metal chips. Look for rock fissures where thin flat layers of rock containing lichens may be pried loose.

Preservation of Lichens

Bulk Methods.   Large quantities of lichens are often preserved in bulk for classroom study or when all of the lichen forms from one ecological habitat are being collected for comparison with those of other such habitats. For these objectives, use bulk preservation techniques. Wash dirt, insects, and other debris from the lichens and sort out those specimens to be preserved. Wrap the entire collection, along with a label containing the field data, in a large piece of cheesecloth and tie with a piece of string. Dry the specimens in a warm room and then store them in closed cardboard boxes which include paradichlorobenzene or naphthalene to control insect pests. One of the nice things about lichens is that the form and color will be well preserved by this method. When ready to study the specimens, soak the entire collection, cheesecloth and all, and the lichens will soon appear amazingly life-like and fresh. They may be redried and remoistened as frequently as necessary.

Museum Techniques.   Lichens are stored in special envelopes which, in turn, are kept in convenient boxes or pinned to herbarium sheets. Dry foliose and fruticose lichens and place them in labeled envelopes. If these are bulky, place them between newspapers and lightly press by setting a book or board on top of a small stack. Never subject lichens to heavy pressure as the resulting distortion of the specimen will render it of less value. Small specimens may be glued to pieces of herbarium sheet cut to fit inside the storage envelope. Crustose forms growing on bark, branches, rock, and so on, may also be glued to small squares of herbarium sheet and kept in envelopes, unless they are too bulky. If the bulk makes some specimens inconvenient, put them in small boxes, along with their data, and keep these in the herbarium cabinet.

The method of folding envelopes and making standard storage boxes and other equipment used for lichens is described in Chapter 4.

To study dried lichens, simply soak them in water for about ½ hour. The specimens will absorb sufficient water to regain their bulk and may be sectioned or otherwise studied. Be sure to fumigate and maintain small quantities of paradichlorobenzene or naphthalene flakes with lichen specimens, as they are subject to attack by insect pests.

MUSHROOMS
AND OTHER HIGHER FUNGI

This section deals with those plants commonly referred to as mushrooms, coral mushrooms, shelf fungi, bird's nest fungi, puff balls, and the like (Fig. 3–1). For the most part, these forms are merely fruiting bodies, or reproductive bodies, which appear seasonally as offshoots projecting from the main body of the plant (mycelium) which grows within soil, dead wood, and so on.

Taxonomic Features

Unlike the woody species of the higher fungi, which undergo little distortion as a result of drying, the fleshy members of this group may suffer great morphological changes and color alteration due to drying. It is essential, therefore, to make notes concerning the morphology and color of the fungi as well as field notes concerning the locality of the collection.

Morphology. The morphology of the common mushroom form (Fig. 3–1A) consists of the cap (pileus), stem (stipe), and the gills which bear spore-producing bodies. In addition, an annulus, or ring, may girdle the stem and a volva, a sac-like membrane, may receive the base of the stem. After the mushroom has been carefully extracted from its substrate, field

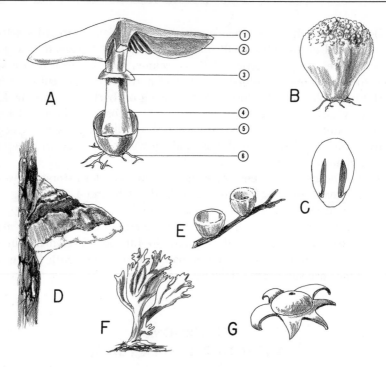

FIG. 3–1. Morphology of higher fungi. A. Mushroom morphology. (1) Cap, (2) gill, (3) annulus, (4) stem, (5) volva, (6) "roots" or rhizoids. B. Puffball. C. Cross-section through a young mushroom. D. Bracket fungus on tree bark. E. Bird's-nest fungi. F. Coral fungus. G. *Geastor.*

notes should at least record the presence or absence of the typical body structures and describe the attitude of the plant itself. Sketches may be essential to denote the position of the cap, with reference to the stem of the more delicate species. The cap may be turned down, directed outward or even directed upward like the walls of a funnel. The manner in which the gills attach to the cap and/or the stem, the position of the annulus with reference to the length of the stem, and the overall shape of the basal portion of the stem where it projects into its substratum must also be noted. The surface texture of the cap and stem may vary from waxy to coarse or porous. Furthermore, it may be smooth or bare, have hair-like, flake-like, or scab-like projections, or conversely, be sculptured and eroded in various patterns. For an excellent series of diagrams relative to these and other characteristics, see Snell and Dick, *A Glossary of Mycology* (1957).

Color. Almost all identification of mushrooms will involve color; hence, accurate records on that subject must be kept. The color of both young and mature specimens should be recorded (see Chapter 1) by means of published color standards, by special color standards presented in some mushroom books, or by recording colors with water-color paints or colored pencils. Note carefully the color of the stem, cap, gills, and spores. The gills change color drastically with age, and may produce spores that appear in a color other than that of the gill. Therefore, spore prints must be made, as described below.

Other Characteristics.   Such characteristics as taste and odor, the presence of a milky sap, and the ability to glow in the dark occasionally prove of taxonomic importance. Thomas (1948) describes taste as "acrid (biting, peppery); astringent (puckery, bitter); disagreeable but not acrid; mild; sweetish." Professional collectors generally record these features in their field notes at the time of collecting (using caution not to swallow any portion of mushrooms being tasted). When describing the locality of a collection be sure to indicate the surroundings of the mushroom, as to whether they were dark and dank or exposed and well lighted. The nature of the substratum supporting the specimen should also be recorded, as well as whether the specimen grew in large clusters, in rings, or in solitude. The season and prevailing weather conditions may also be of some importance.

Collecting the Higher Fungi

Depending on the locality, mushrooms and other higher fungi may appear continuously all year long, although spring, summer and autumn will produce the greatest number of species. Careful records should be kept concerning the weather conditions preceding and during the appearance of different species in order to record the seasonal succession of the fungi.

A box or picnic basket should be used as a collecting receptacle (Burt, 1898; Thomas, 1948; Smith, 1949), as the fleshy species will be crushed and destroyed if they are collected in a plastic bag or similar receptacle. In addition to a basket or cardboard box with rope handles, take along some smaller boxes in which to place delicate forms and some soft paper and string for wrapping individual collections. Paper toweling or premoistened newspaper may be quite suitable for this purpose. Burt (1898) recommends recording field data on small slips of paper and including these in each bundle in order to keep one collection separated from another. Use a large knife or trowel for digging specimens from the soil. A wood chisel or knife may be used for removing specimens from bark and other wooden substrates. Be careful to collect all of the structures surrounding the base of the mushroom where it enters the soil; do not cut the stem at the ground line. Collect large series and attempt to get all stages of development from the button stage on up to fully mature specimens. Spore prints (discussed below) may be initiated at the time of collecting or in the laboratory. If there is any delay in field work or in processing specimens in the laboratory, keep the specimens cool, but do not freeze them. Freezing distorts the tissue and renders the specimens unusable. Finally, record color notes and other field notes before leaving the field. Do not rely on your memory for field data.

Preservation

Remarks. Woody specimens present no real problems in preservation; they are simply dried, fumigated, and stored in boxes or other suitable containers along with their field data and identification slips. Fleshy fungi, on the other hand, cannot be satisfactorily preserved. In liquid preservation (with FAA or 10-percent formalin) the form is retained, while overall shrinking occurs. However, the color quickly disappears and, ultimately, the flesh becomes macerated and will break down. On the other hand, heat drying will retain many of the colors, although considerable shrinking and distortion will follow. Heat drying, nevertheless, is quite permanent and proves to be the only suitable method for museum collections. Frozen specimens are almost always reduced to a disintegrated mess upon thawing. Freeze-drying techniques (Knudsen, 1966) are suitable for museum displays, but render specimens too fragile for taxonomic work.

Spore Prints. Spore prints should always be made of specimens intended for taxonomic work. Select mature specimens with expanded caps and cut off the stem as close to the cap as possible (Fig. 3–2A). Place the cap, gill-side down on a piece of white paper (index card or herbarium paper) and cover this with an inverted glass or dish (Fig. 3–2B). The dish will prevent drafts from blowing spores as they fall during the 6- to 12-hour waiting period. The walls of the gills are lined wtih minute cells (basidia) which in turn produce the colored spores (Fig. 3–2C). If the gills appear whitish or very light in color put a piece of black paper under one quarter or one half of the gill area (Fig. 3–2D); an accurate record may be obtained in this fashion. When the cap is removed, apply several thin coats of artist's spray fixative (used for pencil and charcoal drawings) as directed on the can. This will protect the spores. The print then is included in the packet or herbarium sheet with the mushroom specimens. Figure 3–2E shows an actual spore print made in this manner.

Burt (1898) recommends making spore prints in the field; his technique is slightly modified here. Place the freshly cut cap on a long slip of paper

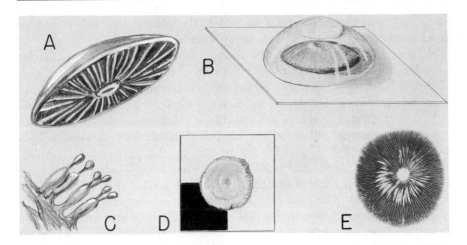

FIG. 3–2.   Spore print technique. A–D. See text. E. An actual spore print of a mushroom.

large enough to go under and over the top of the cap. Slide the "sandwich" into a small plastic bag (freezer or sandwich bag) and secure with paper clips. Place the mushroom cap gill-side down in the collecting basket. By this method spore prints will be ready shortly after you have returned to the laboratory.

Liquid Preservation.   Liquid preservation is suitable when specimens are to be held briefly for classroom use, or for liquid displays when these specimens may be replaced periodically. FAA is a suitable preservative for such use, although no preservative is adequate for long-term use. Research is necessary in this area to determine if some other solution might be used which will maintain specimens in a firm condition and, if possible, retain their color.

Drying Techniques.   Mushrooms are best dried for permanent storage. Drying in sunlight or at room temperature is not satisfactory (Burt, 1898) because decay and the attack of insect larvae will be extensive before the specimens are cured. Specimens should be placed on screen trays or shelves made with ⅛-inch screen or window screen. The screen trays are placed above some source of heat, such as a hotplate or other electrical heating element at a distance where the heat will not burn or destroy the specimens. Dr. Stuntz (in personal communication at the University of Washington) says that heat over 40° C. damages the tissues of mushrooms for taxonomic work. He recommends that temperatures between 35° and 37° C. be used for drying, and that a dilute, 3-percent KOH solution be used to moisten dried specimens when tissue sections are required.

Methods for field drying may be adapted from those outlined in Chapter 5, under "Field Drying Techniques." A very convenient drying chamber is described for the starfishes in the chapter on echinoderms.

Prepare specimens for drying by carefully cleaning away dirt and debris and placing them directly on screens. Place a data sheet nearby, or attach it to the specimens with a string. Large specimens will have to be split lengthwise (Fig. 3–3, A and B) and spread on the screens to dry. Mushrooms are usually fragile and resist bending when first collected, but become pliant during the process of drying (Burt, 1898), so that they can be placed in a desirable way. The caps may be bent down on the stem or spread in a lateral fashion, as shown in Fig. 3–3, C and D. If placed in a moisture chamber or near an open window, mushroom specimens will quickly re-absorb enough water so that final positioning may be completed. Mushrooms must never be pressed, but they may be held flat with a minimum amount of pressure during the final drying process.

Small specimens which are properly placed may be kept in envelopes (described in Chapter 4) and stored in shoe boxes or attached to herbarium sheets. Large specimens, along with their field data, should be placed in small open boxes and kept in trays within the herbarium cabinet. Large woody and fleshy mushrooms which have been dried should be subjected to fumigation to destroy any insects. Place specimens in an airtight cabinet for 24 hours, along with a dish containing a small quantity of carbon bisulfide. Fumigation can also be accomplished by using excessive amounts of paradichlorobenzene for three or four days. Add small quantities of paradichlorobenzene to all fungi collections in the herbarium cases. The woody fungi are most attractive to insect pests and should be observed during biannual fumigation to ensure that pests have not invaded the collection.

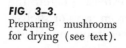

FIG. 3–3.
Preparing mushrooms
for drying (see text).

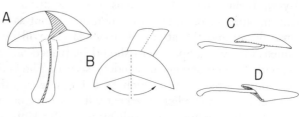

REFERENCES

Alexopoulos, C. J., 1962, *Introductory Mycology*, Wiley, New York.

Barnett, H. L., and Barry Hunter, 1971, *Illustrated Genera of Imperfect Fungi*, Burgess, Minneapolis.

Bessey, E. A., 1964, *Morphology and Taxonomy of Fungi*, Hafner, Darien, Conn.

Burt, E. A., 1898, On Collecting and Preparing Fleshy Fungi for the Herbarium, *Bot. Gaz.* **25**:172–187.

Christensen, C. M., 1965, *Common Fleshy Fungi*, Burgess, Minneapolis.

Christensen, C. M., 1969, *Common Edible Mushrooms*, Univ. of Minnesota, Minneapolis.

Eifert, V., 1952, *Exploring for Mushrooms*, Illinois State Museum, Springfield.

Fitzpatrick, H., and W. Ray, 1955, Some Common Edible and Poisonous Mushrooms, *Cornell Extension Bull.*, 386, Ithaca, N.Y.

Funder, S., 1968, *Practical Mycology Manual for Identification of Fungi*, Hafner, Darien, Conn.

Hedgcock, G. G., and P. Spaulding, 1906, A New Method of Mounting Fungi Grown in Cultures for the Herbarium, *J. Mycol.* **12**:147.

Knudsen, J. W., 1966, *Biological Techniques: Collecting, Preserving, and Illustrating Plants and Animals*, Harper & Row, New York.

Krieger, L. C. C., 1947, *Mushroom Handbook*, Peter Smith, Gloucester, Mass.

Nearing, G. G., 1947, *The Lichen Book*, published by the author, Ridgwood, N.Y.

Ramsbottom, J., 1953, *Mushrooms and Toadstools*, Collins, New York.

Sarles, W. B., *et al.*, 1956, *Microbiology: General and Applied*, Harper & Row, New York.

Smith, A. H., 1949, *Mushrooms in Their Native Habitats*, Sawyer's Inc., New York.

Smith, A. H., 1951, *Puffballs and Their Allies in Michigan*, Univ. of Michigan Press, Ann Arbor.

Smith, G. M., 1955, *Cryptogamic Botany, Vol. I: Algae and Fungi*, McGraw-Hill, New York.

Snell, H., and E. A. Dick, 1957, *A Glossary of Mycology*, Harvard Univ. Press, Cambridge, Mass.

Stanier, R. Y., *et al.*, 1970, *Microbial World*, Prentice-Hall, Englewood Cliffs, N.J.

Thomas, W., 1948, *Fieldbook of Common Mushrooms*, Putnam, New York.

The Mosses and Liverworts

This chapter will include a brief discussion on the so-called "club mosses" (Tracheophyta), in addition to the true mosses, because methods of preparation are the same.

THE MOSSES

The true mosses, found the world over, are moisture-loving plants which possess root-like, stem-like, and leaf-like structures in the gametophyte generation. Students should consult *Biological Techniques* (Knudsen, 1966), Conard (1956, an excellent beginning source), or other works to acquaint themselves with moss structures and life history stages, which are essential to their identification and study. It is necessary to collect mature gametophytes, with their gametangia well developed, in one season of the year, and to collect specimens in which the capsule of the sporophyte is at a point of maturity during the opposite season. The same is true of the club mosses, for specimens gathered without the fruiting cones will be of less value taxonomically.

Collecting Methods

Provide yourself with a vasculum, a large plastic bag, or knapsack in which to carry specimens. Use small plastic freezer bags or sandwich bags and rubber bands to isolate individual specimens, along with their field data. Take along a pocketknife and a trowel for dislodging specimens attached to stones, bark, and the like, or for digging specimens from the soil. Be careful to dig up specimens so as to obtain the root-like rhizoids. Select only specimens that are mature and which possess reproductive structures. Several field trips throughout the year may be essential if the collector is to obtain local specimens in peak condition. Collect a large number of specimens whenever possible. Carefully wash mud and debris away from the rhizoids and basal portions of the gametophyte. Specimens that grow upon rocks, bark, and so on, have a habit of collecting soil which should be washed away. When collecting free-floating mosses from ponds, select specimens that are actively growing rather than dead or dormant specimens. Isolate each collection and insert a field note and number referring to the collecting site and data in the field notebook.

Preservation of Specimens

Arrangement and "Pressing."   Specimens to be stored in envelopes or on herbarium sheets are carefully washed, blotted partially dry, and arranged on several thicknesses of newspaper folded in quarters. Position the specimens to display all of the essential body parts to their best advantage. Highly branched forms must be spread out, to avoid multiple overlapping at any one point. Conard (1945) recommends stacking papers with specimens in piles of about 12 inches each, and then adding a light weight of about 2 pounds to "press" the specimens. Do not add enough weight to literally "press" the specimen, but only enough to keep it in a two-dimensional plane. When leaflets and other structures are flattened against the main body of the plant the taxonomic characteristics are difficult to analyze. If specimens become misshapen they may be dipped in water for a few moments, until they are pliable, and then "repressed" in the desired posture.

Envelope Storage Technique

Most collections of mosses are kept in envelopes or packets, along with their field data. These envelopes are in turn stored in boxes 4 by 6 by 17 inches, or they are attached to herbarium sheets.

Select a good grade of 20-pound bond paper with at least 50-percent rag content, in the standard 8½- by 11-inch size. If boxes (shoe boxes or the like) are used for storage, determine an envelope size that will conveniently fit into them. Standard envelopes are usually 5½ inches wide and from 3¾ to 4½ inches high. An older style of envelope is made by following Fig. 4–1, steps A, B, C. For better envelopes, follow Fig. 4–1, steps A, D, E.

Write the scientific and common name of the specimen, along with its field data, on the upper flap of the envelope. (Consult Chapter 5 for field label data and techniques, as well as the rubber-stamp technique, which should prove useful in this instance.) Half-size envelopes are frequently made to hold very small specimens.

Herbarium Sheet Mounts

A second common practice is to mount moss specimens directly on herbarium sheets. For this technique follow the instructions in Chapter 5. The glue-plate method will probably prove most suitable. Direct mounts have

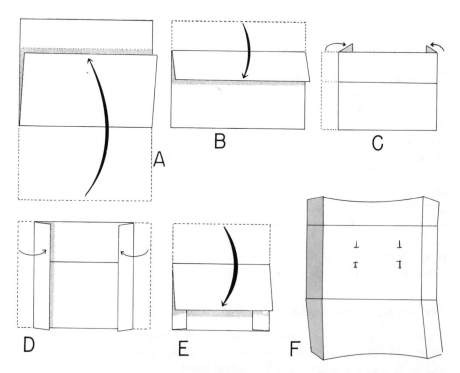

FIG. 4–1. Two methods of folding envelopes for storing all types of botanical specimens, either in filing boxes or by pinning them directly to herbarium sheets (see text).

the advantages of making specimens more easily accessible and more readily viewed. However, it is impossible to adequately study such specimens under the microscope or to boil or steam them in order to manipulate the various parts of the foliage. Because of their fragile nature mosses must be protected in genus folders (Chapter 5).

Bulk-Dry Methods

If it is desirable to preserve large quantities of mosses for classroom observation, or to preserve the larger club mosses in a natural posture, the bulk-dry method will be convenient. Mosses are cleaned in the usual manner and carefully arranged on a piece of cheesecloth, along with a slip containing the field data. The cloth is then wrapped in such a manner as to contain the moss but not to distort its shape. Tie the corners of the cheesecloth together or secure the bundle with string. Allow the specimens to dry and store them in a pasteboard box with some paradichlorobenzene. For larger club mosses record the collecting data and scientific name on a waterproof label (1 inch by 2 inches) and tie this to the stem of the fresh moss. Moss specimens so prepared should be carefully placed in a cardboard box and permitted to dry. When they are dry, some paradichlorobenzene is added and the box is tied shut and stored in a dark place. These dry specimens are quite adequate for casual observation or they may be subjected to soaking, steaming, or even dipping in boiling water, to permit their manipulation in the classroom.

Liquid Preservation

Moss capsules and other structures preserved at their prime in liquid are useful for taxonomic or classroom dissections. For normal preservation use FAA. Chapter 5 describes a solution that will to some extent preserve the natural green color of the plant, if this is desired.

LIVERWORTS
AND HORNWORTS

At first glance liverworts are inconspicuous plants produced as a flat thallus or as a central axis with simple or multilobed leaves. Hornworts resemble the flattened thallus of the liverworts, but tend to produce erect, spindle-shaped sporophytes from which the common name is derived. As in the cycle of the mosses, the gametophyte generation of liverworts and hornworts is most conspicuous, whereas the sporophyte generation tends to be less conspicuous in appearance. Alternation between the asexual and sexual generations is the common mode of reproduction. However, simple asexual reproduction by budding within the gemmae cups or by fragmentation is also common. Most of the liverworts possess root-like rhizoids, which are produced from single cells; these provide a means of attachment to the substrate. Both generations are photosynthetic, and thus have a lusty green color.

Habitat and Collecting

Liverworts prefer moist, shaded areas usually associated with some source of water. Stream and lake banks, moist road banks in forested areas, and similar habitats, will support the terrestrial species. Look carefully among grasses and other vegetation for liverworts growing in any of these situations. Where the ground is moist and damp, lift up bushes and overhanging branches to inspect for liverworts. Ponds and shallow lakes are a favorite habitat for the aquatic species. One worker suggests that hornworts require a temperate microclimate with acidic substrates mostly of soil, but occasionally are found on bark.

Use plastic freezer bags of about one-quart capacity for collecting. A knife, trowel, or dip net (see Chapter 12 for construction) is used for removing specimens from their substrate or dipping them from water. Place the specimens in plastic bags, inflate the bag, then seal it with a rubber band. When placed in a carrying box, the inflated bag will prevent crushing when other specimens are added. Take careful field notes concerning the exact microhabitat, dominant species of plants associated with the habitat, and light and moisture conditions. Include a copy of this information or a reference number to these data in the field notebook in each plastic bag.

Collect well-formed, mature specimens showing ripened reproductive bodies whenever possible. Liverworts are easily kept alive in the aquarium or laboratory terrarium and may be cultured until reproductive stages appear and mature. This is the only suitable way of getting excellent specimens, unless numerous field trips can be made. If kept cool and provided with some moisture, specimens will last many days in their plastic bag containers and can be transported from very long distances. Provide an aquarium with pond water for aquatic species; for terrestrial species provide a sloping surface in a terrarium containing a small quantity of water. A mixture of peat moss and loam makes an excellent substrate in which to plant liverworts. Place a sheet of glass over the aquarium or terrarium to maintain the humidity level, and keep the specimens in somewhat subdued light such as that from a north window.

Preservation Techniques

Liverworts should be pressed between specimen sheets and dryers with only moderate pressure, following the techniques described in Chapter 5. Should specimens tend to stick to the specimen sheets, place them directly on herbarium paper and cover them with cloth, as described for the marine algae. Either the gummed-cloth or herbarium-paste method (both discussed in Chapter 5) should be used for attaching specimens to standard herbarium sheets. The herbarium label, containing the scientific name and field data, is attached to the lower right-hand corner of the sheet. Herbarium sheets are stored in genus folders. An alternate (and preferable) method is to store the pressed material in moss envelopes, as described above. In pressing, be sure to apply only enough pressure to hold the plant in a two-dimensional plane, but not enough to crush it or totally flatten it.

Liquid Preservation

For many purposes liquid preservation is quite suitable. It is advisable to preserve some specimens in liquid, even though the bulk of these are to be dried for herbarium sheets or envelopes. Several solutions are quite suitable for this work. (1) A 4-percent formalin solution, with or without a small quantity of copper sulfate, should be used for general preservation. (2) FAA (see Appendix C) is very suitable for general preservation and should be used for specimens intended for microscope slides. Again, a small quantity of copper sulfate may be added if desired. (3) A more complex formula for maintaining the green coloration of plants is presented in Chapter 5.

REFERENCES

Bodenberg, E. T., 1954, *Mosses: A New Approach to the Identification of Common Species*, Burgess, Minneapolis.

Chamberlain, Edward B., 1903, Mounting Moss Specimens, *The Biologist* 6:102–103.

Conard, H. S., 1945, The Bryophyte Herbarium. A Moss Collection: Preparation and Care, *The Bryologist* 48(4):198–202.

Conard, H. S., 1956, *How to Know the Mosses and Liverworts,* Brown, Dubuque, Iowa.

Grout, A. J., 1965, *Mosses with Hand Lens and Microscope,* Lundberg, Ashton, Md.

Jennings, O. E., 1951, *Manual of the Mosses of Western Pennsylvania and Adjacent Regions,* Univ. of Notre Dame Press, Notre Dame, Ind.

Knudsen, J. W., 1966, *Biological Techniques: Collecting, Preserving, and Illustrating Plants and Animals,* Harper & Row, New York.

CHAPTER 5

The Higher Plants
(Tracheophyta)

The phylum Tracheophyta contains a number of subphyla with diverse species which are all related by the common characteristic of having tracheids (a type of vascular tissue) or tracheid derivatives. To most of us the tracheophytes are the most familiar and important plants. Some of the plant groups involved, their characteristics, and the techniques required for their study will be discussed.

SUBPHYLA
OF THE HIGHER PLANTS

Club Mosses

The club mosses (Lycopsida) resemble the true mosses and some of the gymnosperms, or conifers, in both appearance and foliage. The plant consists of a rhizome (Fig. 5–1) which bears roots and gives rise to erect stems. In season the erect stems will bear cones, or strobili. These cones bear both kinds of sporangia and thus, popularly but incorrectly, are both "male" and "female." The entire plant, including rhizomes and roots, should be collected and dried in the manner of true mosses (see Chapter 4).

Horsetails

The horsetails, subphylum Sphenopsida, are primitive plants that are widely distributed. They grow profusely along some highways, on wet banks near lakes, and in moist forested areas. Horsetails somewhat resemble the ferns in their gross structure and life history. The plant consists of a rhizome which gives rise to roots and to stems bearing needle-like leaves. Some of the stems give rise to spore-bearing cones. When released, the spores develop into the bisexual gametophyte generation. Collect all plant parts and, if necessary, make special trips to get immature and mature cones. The gametophytes may be reared in the same manner as fern prothallia.

Ferns

Ferns represent one large class of plants belonging to the subphylum Pteropsida. This subphylum also includes the conifers and the true flowering plants. In collecting ferns, gather more than just the frond, or leaf (see Fig. 5–1, A, B, and C); get the entire stipe, the rhizome, and some of the roots. Collect some fronds bearing reproductive bodies (sporangia) and mount these so that the sporangia may be seen on the herbarium sheet. The sexual, or gametophyte, generation of ferns may be reared on agar, with plant food added, or on clean bricks which are kept moist with a solution of plant food and water. These are best preserved in liquid or on permanent microscope slides.

Conifers

The conifers represent the class Gymnospermae of the subphylum Pteropsida and include such familiar plants as pines, cedars, and firs. In collecting throughout the year obtain both the "male" and "female" cones, as well as needles, terminal branches, growth buds, and the like, from the same plant. Take notes on the nature of the bark, the height of the tree, the altitude, the nature in which the cone comes off the branch, and other things of taxonomic use. Large cones are dried; smaller ones may be preserved in liquid or pressed, along with the main plant specimen.

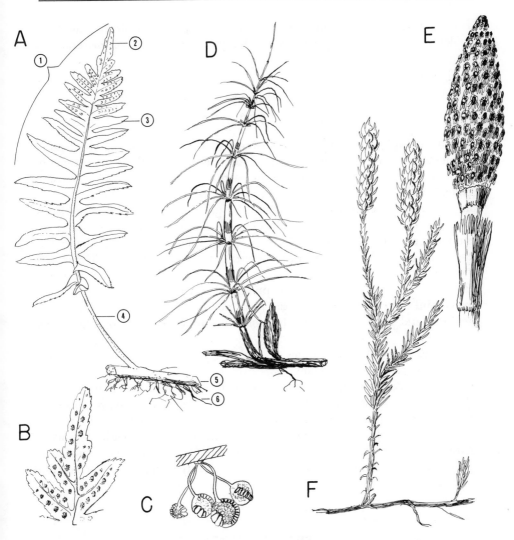

FIG. 5-1. Some higher plants, A–C. Fern morphology. (1) Frond, (2) sporangia, (3) pinna or leaflets, (4) stipe, (5) rhizome, (6) rhizoids or roots. D–E. Horsetails. F. Club moss.

Flowering Plants

The flowering plants, the class Angiospermae, of the subphylum Pteropsida, should be carefully collected so that all plant structures, including fruits and seeds, are obtained. The general description of collecting methods which follows gives many suggestions for the treatment of this material.

PURPOSE
OF THE HERBARIUM

Plant collections maintained by botanists vary in both size and purpose. Students who are interested in plant taxonomy in general, or the taxonomy of some family or genus of plants in particular, may make very extensive collections. When carefully identified and labeled, the specimen becomes an important reference for illustrating the particular species characteristics and for checking future identifications. With reference collections new students may learn the procedures of plant identification and determine the accuracy of their own work by referring to the herbarium collection. In taxonomic collections field data, plant measurements, and plant character-istics are an important part of the preserved specimens. The collector must be careful to get all of the essential plant portions which serve as taxonomic features and to cross reference his collection in such a way that all forms of collections may easily be surveyed. Complete specimens with ample field data lend themselves to studies of ecology, plant life histories, and plant geography, as well as taxonomy.

Many teachers maintain small plant collections of selected material to illustrate such things as different groups of plants, types of leaves, flowers, fruits, seeds, cones, or other features that may be of importance in the teaching of plant sciences. Such collections should be made with the utmost care to provide all of the taxonomic parts required and to demonstrate the proper technique for herbarium specimens. Poorly prepared specimens are of little value in teaching on any level.

FIELD METHODS
AND MATERIALS

Collecting Tools

A number of small tools will be required for field collecting. A pocket-knife, hand pruning shears, field notebook, 6-foot steel tape, some form of trowel, a vasculum, and a plant press make up the essential list. The trowel is used for extracting the roots of small plants; the steel tape may be used to take data on height of the population. The vasculum consists of a metal cylinder with a sliding door, usually worn on a strap over the collector's shoulder, into which plant specimens are placed. However, large plastic bags are more versatile and suitable than the vasculum, for they prevent more fully dehydration of specimens between the time of collecting and the time of pressing. A collector should carry several small plastic or paper bags for fruits, bark, and other specimens.

Methods of Collecting

Collecting does not start in the field. In order to collect usable plant specimens you must know what essential plant parts are needed from each group of plants. Thus a study of plant keys and literature (see the characteristics given above, or Fogg, 1940, and others) will ensure more accurate field collecting.

When collecting small herbs, pick out plants of medium size which show both mature flowers and (if possible) mature fruits (immature flowers and fruits are of little value). Get the complete plant, including the roots, stems, leaves, and floral and fruiting portions. Collect extra flowers and fruits for identification. Carefully dig up the plant and wash the dirt from the roots so that the root mass may be included on the finished pressed specimen. Specimens up to 3 feet in length may be handled in their entirety, provided they are not too thick and woody, by bending the stem once or twice in order to make the plant fit the herbarium sheet. Collect many specimens of smaller plants and several of the medium and larger ones to ensure that all of the essential characteristics will be displayed on the herbarium sheet. Collect within an area repeatedly, if necessary, to obtain mature flowers and fruiting bodies. Take ample field notes and measurements to complete the record.

When collecting larger plants, choose portions with flowers, fruits, and a representative of leaf types. Leaves on the terminal branches of trees often differ from those down low on the trunk, so that several samples must be taken. Samples of bark and fruits may be put into plastic bags, along with a field number, and later may be incorporated with the finished specimen. If you are looking for cones, seeds, or fruits of various plants, make sure those specimens picked up from the ground actually came from the plant in question. In some plants the flowers, leaves, and fruits appear at different times; thus, tag a particular tree or plant and collect from the same specimen throughout the year until all of the essential parts are obtained. Higher branches may be collected with the aid of a 40- or 50-foot rope. Fill a small cloth bag with dirt and tie this securely with a piece of string. Tie the end of the string to the rope in such manner that if the bag becomes entangled in the tree the string will break and release the rope. Next, carefully toss the weighted end of the rope into the outer branches of the tree and, when the rope is entangled, pull the specimen down from the tree.

When collecting club mosses, horsetails, or ferns make sure that root specimens, complete stems, and fruiting bodies, as well as sterile portions, are collected.

Put specimens into the press as quickly as possible, using the technique described below, making sure that the field data or a field number is placed on each specimen sheet. Coarse brush that will not quickly wilt may be kept in the order of collection by attachment to a stout cord with slip knots.

In the tropics the almost constant rain and extreme humidity make the drying of specimens and the maintenance of specimens almost impossible without artificial heat and other facilities. Schultes (1947) found that collecting by normal methods in the Amazon drainage area of Colombia was impossible and resorted to temporary preservation of plant specimens. His method consisted of pressing plants for 24 hours between sheets of folded newpaper, after which they were dipped for a few moments in a strong formaldehyde solution consisting of two parts of commercial 40-percent formaldehyde and three parts of water. The dripping specimens were, then replaced in their newspaper folders and put back into the press. The specimens, press and all, were then wrapped in an airtight rubberized bag and shipped to a laboratory where the plants were pressed and dried in the conventional manner. Schultes found that when shipping time was less than 15 days a solution of one part full-strength formaldehyde to two parts water was strong enough to prevent rotting. Moore (1950) reports excellent results from preserving bundles of plants in a 1-percent solution of hydroxyquinoline sulfate, presumably in plastic bags, and then later pressing the specimens. He also found that palm fruits and flowers could be preserved in the same concentration and transported to the museum in that way.

The Plant Press

The plant press is an indispensable tool by means of which fresh plant specimens are pressed and dried in preparation for permanent mounting. The many styles of plant presses fall into two chief categories, the field press and the laboratory press. These are available through biological supply houses, but a student can make his own press at very little cost (Fig. 5–2).

Using the Plant Press.   Obtain some form of botanical dryers which absorb moisture from the plant specimens, and some ventilators which permit the circulation of air between each group of plants being pressed. A

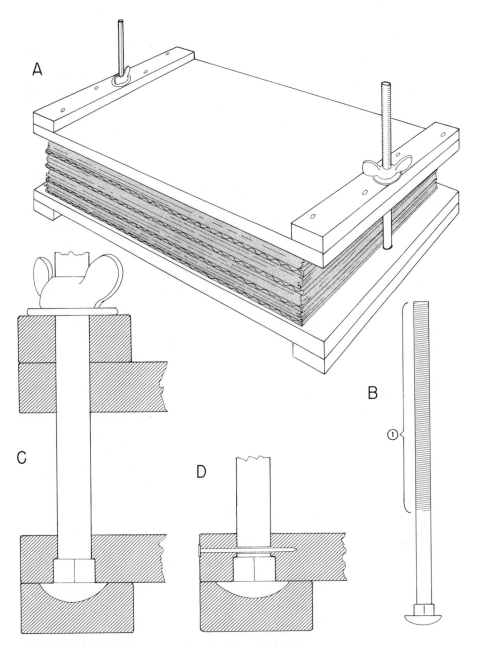

FIG. 5–2. Methods for building a laboratory plant press using ¾- or 1-inch plywood, 12 by 21 inches in size, and two ½- by 12-inch carriage bolts (B, 1) threaded 8 inches down. A ⅛-inch hole drilled after the bolt is in place receives a nail which prevents turning. See *Biological Techniques* (Knudsen, 1966) for full instructions.

fine-grained absorbent felt dryer or blotter, 12 by 18 inches, may be purchased, but folded newspaper is just as satisfactory. Likewise, commercial corrugated cardboard ventilators or aluminum ventilators, 12 by 18 inches, may be purchased but can easily be made from large pasteboard boxes. The corrugated ventilators permit air to circulate between the dryers and draw moisture off more readily. Finally, provide specimen papers made from single sheets of newspaper folded over.

Put a stack of ventilators, dryers, and specimen papers in your press when going into the field. Press specimens shortly after collecting, so that wilting will not distort their appearance. After a specimen has been collected, put a ventilator and a dryer, in that order, on the bottom board of the press. Open a specimen sheet and put this on top of the dryer. Arrange the plant in the way you wish the finished specimen to appear, remembering to show all of the floral parts, leaves, stems, and roots, so that every taxonomic feature can be observed. Arrange the roots on the lower part of the folder and, if necessary, bend the stem so that the bulk of the plant is distributed uniformly. Do not let several parts overlap, causing unnecessary thickness in any one area. Show both sides of leaves and both sides of large disc-shaped flowers such as daisies. When the corolla is tubular in nature, split the tube of one or two flowers to show the inner parts. Add extra flowers and fruits that may later be stored in a paper envelope attached to the herbarium sheet. These will be useful for dissection and microscopic study. On the inside of the specimen sheet write the field data or, at least, a field number that corresponds to data kept in the notebook. The specimen sheet will remain with the plant until it is finally mounted on a herbarium sheet. Hold the plant in place and fold over the other half of the specimen sheet, add another dryer and a ventilator, and, finally, put the press together again and retighten the straps.

When the press is full of specimens, place it on the ground and, while kneeling on it, tighten the press straps. Change the dryers every 12 to 24 hours to reduce drying time and prevent molding. Wet dryers must be placed in the sun or dried artificially so that they will be ready for reuse. When the dryers are changed in the laboratory, specimens may be transferred from the field press to the laboratory press, if this is desirable. When specimens are dry, stack them in their specimen sheets, between two ventilators, and tie them in bundles of about 50 plants. These are now ready for identification and mounting.

When working with thick succulent plants such as the cactus, halve the flat segments lengthwise and scoop out the inside so that only the surface is pressed. Cylindrical sections may be cut to appropriate lengths, split down on one side and hollowed out, rolled out flat, and pressed. The tops of small barrel cacti should be cut off, with a minimum of flesh attached, scooped out, and pressed as a disc. The side of the barrel is treated as a cylinder, split, cleaned, and flattened. Large flowers may be cut in half lengthwise and the fleshy base reduced before pressing. Otherwise, the receptacle may be cut off just below the petals and the flower pressed with all the petals radiating like those of a daisy. Occasionally, the folding of the spines will become difficult, but this may be achieved by putting a dryer and a ventilator board over the cactus and simultaneously pressing downward and to one side. This motion will fold the spines over and permit the bulk of the cactus to be reduced.

Some flowers, such as the iris and many aquatic forms, will stick to the specimen papers or dryers. In such cases the flowers may be placed directly on a herbarium sheet and covered with waxed paper or very fine cotton cloth. After drying in the usual manner, the waxed paper or cotton cloth is very carefully peeled away. The stem and leaves must be strapped down after pressing so that the flower is not broken.

Many of the conifers lose their needles, even though they may be covered and given the best of care. Sharp (1935) recommends a method which is slightly modified here. Cut fresh specimens to fit the herbarium sheet and then put them on a plate of glass covered with partially diluted herbarium glue (see mounting methods below). Carefully lift the specimens and put them on the herbarium sheet, being careful not to have an excess of glue. Cover these with several dryers and press until dry. Another method, recommended by Wherry (1949), takes advantage of one of the clear plastic sprays, sold under the name of Kraylon. He recommends that the conifer be dried in the usual manner and then sprayed where the needles attach to the stem. The specimen is then mounted on a herbarium sheet and may be covered with cellophane, although this is unnecessary.

Field Drying Methods.   It is often imperative to hasten the drying of pressed material, both to free the presses for more specimens and to prevent them from becoming moldy or rotten. In sunny weather field presses (Fig. 5–3) may be set in the sunshine on pavement or hot sand, or suspended by screw eyes in the sunshine. In very damp climates, however, it may be essential to use artificial heat for drying. Specimens that are carefully dried with artificial heat, not exceeding 115° F., will show no appreciable difference in appearance from those dried normally. If electricity is available in the field, place your specimens in a dryer constructed of a pasteboard box, a light bulb and receptacle, and an extension cord, as described in Chapter 12 under "Field Drying." Place the plant presses on edge over the source of heat so that hot air will continuously circulate through the ventilators and remove moist air. Gates (1950) describes in detail the construction of a

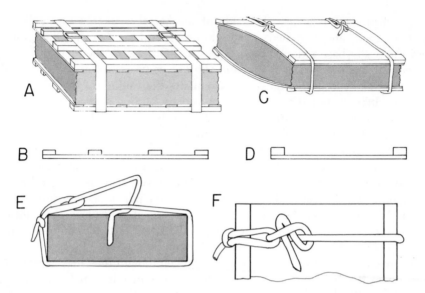

FIG. 5–3. Field presses. A–B. Lath construction. C–D. Plywood construction. E–F. Method of tying a field press with rope. Field presses measure 12 by 18 inches. See *Biological Techniques* (Knudsen, 1966) for full instructions.

more elaborate electrical dryer for herbarium specimens designed primarily for museum use. However, an inexpensive controlled-heat dryer may be made by providing a box, over which the presses may be placed, and providing this with a coil-heating element or a bank of electric lights and a chick incubator thermostat (which costs around one dollar).

When electricity is not available, the heat from pressure gasoline lanterns or stoves, or from kerosene lanterns, serves quite well. Place the plant presses on edge on some sort of rack or suspend them by means of rope and screw eyes over a Coleman lantern and drape a sheet of muslin skirt around the presses to funnel the hot air up through the ventilators. In using this technique be extremely careful to provide space for air to enter underneath the sheet and keep the plant presses far enough away from the lantern to prevent burning. If kerosene lanterns are used, be sure the fuel is clean; otherwise, soot will be deposited throughout the ventilators.

Field Notes and Data

Students should get into the habit of taking field notebooks with them and recording accurate data while collecting. Although the amount of field data recorded may depend on the intended use of the specimens, only those specimens which have complete data will be of universal value in botany. The recognition of new taxonomic characteristics often makes the splitting of species into respective subspecies impossible unless accurate notes have been kept. Keeping of locality data, arranged according to serial numbers, is described in Chapter 1. In addition to locality data, however, the student should record information concerning the species itself and its surroundings: the type of plant association (meadow, woodland, and so on), associated plants (sagebrush, etc.), altitude, the slope of the ground, the nature of the soil and its moisture content, and other pertinent factors. Insofar as the species itself is concerned, record the average size of specimens, the color and pattern of the flower, the nature of the bark for trees, and any other detail that the preserved specimen will not give after color has faded. The recording of actual color notes for new species, or for species whose identification depends on color, is discussed in Chapter 1. Color photographs are not very suitable, unless some color standard is included in the picture to serve later as a norm in the laboratory.

IDENTIFICATION, MOUNTING, AND PRESERVATION

Identification

Whenever possible, collect extra material for immediate identification. Transfer species names to the proper specimen sheets during the first change of dryers on the plant press. If immediate identification is not possible, identify specimens prior to mounting them on herbarium sheets. One or two flowers may be removed from the specimen for dissection if extra flowers and fruits have not been pressed. Place the flowers in a bath of boiling water from one to three minutes or hold them over the spout of a steam kettle until they are thoroughly relaxed. Return the dissected parts to paper envelopes and affix these to the herbarium sheet along with the main specimen. If extremely large numbers of plants must be handled and mounted, it is best to postpone identification until after mounting.

Mounting Pressed Specimens

Mounting is a process whereby pressed and identified specimens are attached to herbarium sheets for both protection and permanent storage. From one to many specimens representing only one species may be mounted on a single sheet. Arrange the specimens so that the bulk is more or less uniformly distributed around the sheet (Fig. 5–4E). Herbarium paper comes in a standard size, 11½ by 16½ inches; it should be of a good quality, preferably a high-rag-content paper that will not yellow with age. The standard weight approximates that of manila folder stock or Bristol ledger, but heavier 5-ply mounting cards are also obtainable.

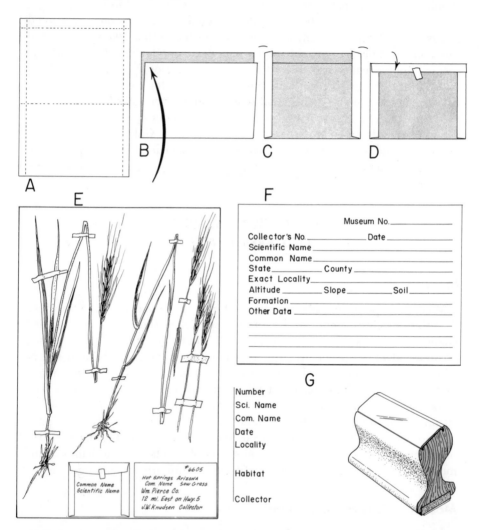

FIG. 5–4. Herbarium sheets and mounting techniques. A–D. Folding a storage envelope for the herbarium sheet. E. Mounting specimens on a herbarium sheet with cloth tape (see text). Note storage envelope and label at the bottom of the herbarium sheet. F. A herbarium label modified from the U.S. Forestry Service label. G. A simple, rubber-stamp label suitable for museum work.

There are several good methods of mounting. An older method still followed utilizes herbarium tape, obtainable from biological supply dealers in rolls from ¾ to 2 inches wide. Make narrow straps of tape by cutting the rolls crosswise. Apply tape straps amply so that the plant will be flexible and yet firmly attached to the sheet. Plastic and celluloid tapes are not suitable, as they are not permanent. Tape may be made by using thin strips of cloth to which herbarium paste has been applied.

Another method consists of spreading diluted Elmer's Glue (equal parts of glue and water) or other glue on a glass sheet and placing the specimen on the glass until the undersurface is thoroughly coated. Then the specimen is applied to the card or the card to the specimen, as described above. This method is extremely quick and very satisfactory. Cloth straps to hold the plant in position will not be necessary unless the glue used is untrustworthy.

The final method consists of using a new plastic mounting medium that takes the place of gluing or strapping, and is extremely useful when large numbers of specimens must be mounted. The mounting material, developed by Archer (1950), is made with the following formula: 800 cubic centimeters of toluene, 200 cubic centimeters of methanol, 250 grams of ethyl cellulose (standard 7 cps.), and 75 grams of Dow Resin 276 V-2. This is mixed to the consistency of a heavy syrup and is applied from a squirt oil can or polyethylene squirt bottle. The material is now available in small quantities from Carolina Biological Supply. In this method specimens are placed on herbarium sheets and, if necessary, weighted down. The medium is squirted over stems or leaves in the form of straps, as in the cloth strap method, in sufficient number to firmly affix a plant. After drying, the plant may be permanently stored.

Liquid Preservation

Preserved specimens are useful in general laboratory studies when fresh material is unavailable and dried specimens are unsuitable.

Preservation for General Use.   When flowers, fruits, stems, leaves, and other plant parts are required for classroom study, preservation in a 4-percent formalin solution is satisfactory. For large, fleshy specimens, or when many specimens are placed in one jar, use a 5- to 6-percent solution. To facilitate speedy penetration of plants, especially those with waxy coats, replace from 20 to 50 percent of the water with alcohol. A general disadvantage with the formalin or formal-alcohol preparation is that color will be bleached or removed entirely. Specimens should be soaked in fresh water for several hours prior to classroom use.

Preparation for Chlorophyll Retention.   On many occasions the retention of green color is extremely useful in the study of small organisms such as fern prothallia. A satisfactory solution for retaining color (*Turtox Service Leaflet* 3) is made by adding 20 grams of phenol c.p.; 20 grams lactic acid, specific gravity 1.21; 40 grams of glycerin, specific gravity 1.25; 0.2 grams cupric chloride; 0.2 grams cupric acetate; and 20 cubic centimeters of distilled water. Specimens are stored in this solution until required.

Color Preservation.   No method for the color preservation of flowers or fruits is everywhere satisfactory. Color retention by any method is not wholly satisfactory for taxonomic purposes and, therefore, one should use color notes made at the time of collection (see Chapter 1). Many dried plant specimens will retain natural colors for very long periods of time, but exposure to light, humidity, temperature, chemical action, and other factors may, gradually or quickly, alter color. Scully (1937) developed solutions that give satisfactory results for many flower and fruit colors. To use this method, place specimens in vials containing a 5-percent copper sulfate solution for 24 hours to set the color. Wash the specimens several times and put them into a second solution of 16 cubic centimeters of sulfuric acid, 21 grams of sodium sulfite, and 1000 cubic centimeters of water. The material is maintained in the solution in stoppered vials. According to Scully, delicate pinks and blues or some of the deep reds and purples will fade, whereas many of the colors, especially the yellow range, will show very little alteration.

Preservation for Histology.   When specimens are intended for tissue studies and slide preparation, preservé them in a formal-acetic-alcohol solution made of 6 parts formaldehyde, 50 parts 95-percent ethyl alcohol, 2 parts glacial acetic acid, and 40 parts distilled water. Specimens may be stored in this solution indefinitely for general use or for histological techniques.

LABELS AND STORAGE

Labeling and Labels

Specimens are of little value without labels that reflect the data recorded in the field notebook. Labels for preserved specimens should be written with waterproof India ink, dried for about 10 minutes, and then placed inside the container, along with the specimen. Labels for herbarium sheets, usually attached to the lower right-hand corner, measure about 3 by 5 inches. The label should include: (1) the locality, including state, county, and distance from the nearest town; (2) date; (3) field number or collector's number; (4) scientific name; (5) common name; (6) habitat, which may include the slope, the soil, the altitude, associated vegetation, etc.; (7) collector's name. Such labels are usually printed to the collector's specifications and headed with the name of the institution or the collection represented, as shown in Fig. 5–4F. However, for small collections labels may be made by means of a rubber stamp applied directly to the herbarium sheet (Fig. 5–4G). The following headings should be arranged in a column: locality, date, field number, scientific name, common name, habitat, collector. Be sure to leave sufficient room between headings to permit the transcription of adequate field data. The label is stamped with black ink about 5 inches in from the lower right-hand margin.

Storage Techniques

Problems Involved.   Light, moisture, insect infestation, and rough handling represent the common elements which damage dry, mounted herbarium specimens. Specimens that are properly cared for and handled will not wear out or disintegrate. After identification, file the members of a particular genus in one or more herbarium folders and store these in a herbarium cabinet. The folders are usually made of manila and are slightly larger than the herbarium sheet. When handling individual sheets, be careful not to bend or twist the specimens; pick up each sheet individually and set it aside when working through a stack. With proper handling one need not cover specimens with thin plastic. For general classroom use, however, covering the specimen is often advantageous. Always keep some fumigant such as paradichlorobenzene or naphthalene flakes inside the herbarium cabinet.

Arrange your specimens either alphabetically or taxonomically within the cabinet. The keeping of a card file is not essential for a small collection but is highly desirable in large herbaria. Such a file cross-references the collection and makes the specimens more readily available for study. See *Biological Techniques* (Knudsen, 1966) for methods of constructing herbarium storage cabinets.

All materials preserved in liquid should be kept in the dark. Vials should be numbered and kept in sealed jars containing the same preservative. Type vial numbers or specimen names on herbarium paper and place this label inside the jar for the convenience of relocating specimens.

REFERENCES

Archer, W. Andrew, 1950, New Plastic Aid in Mounting Herbarium Specimens, *Rhodora* 52(623):298–299.

Baerg, Harry, 1955, *How to Know the Western Trees*, Brown, Dubuque, Iowa.

Blackburn, B. C., 1952, *Trees and Shrubs in Eastern North America*, Oxford Univ. Press, New York.

Bowers, N. A., 1965, *Cone-bearing Trees of the Pacific Coast*, Pacific Books, Palo Alto, Calif.

Canada Department of Resources and Development, 1949, Forestry Branch, Native Trees of Canada, *Canada Dept. Res. and Devel. Bull.*, **61.**

Cobb, Boughton, 1956, *A Field Guide to the Ferns*, Houghton Mifflin, Boston.

Durand, H., 1949, *Fieldbook of Common Ferns*, Putnam, New York.

Fernald, M. L., 1950, *Gray's Manual of Botany*, American Book, New York.

Fogg, John M., 1940, Suggestions for Collectors, *Rhodora* 42(497):145–157.

Frye, T. C., 1934, *Ferns of the Northwest*, Binfords & Mort, Portland, Ore.

Gates, Burton N., 1950, An Electrical Drier for Herbarium Specimens, *Rhodora* 52(618):130–135.

Graves, A. H., 1956, *Illustrated Guide to Trees and Shrubs*, Harper & Row, New York.

Harlow, William M., 1942, *Trees of the Eastern United States and Canada*, Dover, New York.

Harrar, Elwood S., and J. George Harrar, 1946, *Guide to Southern Trees*, Peter Smith, Gloucester, Mass.

Jaques, H. E., 1946, *How to Know the Trees*, Brown, Dubuque, Iowa.

Knudsen, J. W., 1966, *Biological Techniques: Collecting, Preserving, and Illustrating Plants and Animals*, Harper & Row, New York.

Matthews, F. S., 1955, *Field Book of American Wild Flowers*, Putnam, New York.

Moore, H. E., Jr., 1950, A Substitute for Formaldehyde and Alcohol in Plant Collecting, *Rhodora* 54(653):123–124.

Parsons, Frances Theodora, 1961, *How to Know the Ferns*, Dover, New York.

Preston, R. J., Jr., 1961, *North American Trees*, Iowa State Univ. Press, Ames, Iowa. (Available in paperback from M.I.T. Press, Cambridge, Mass.)

Schultes, Richard E., 1947, The Use of Formaldehyde in Plant Collecting, *Rhodora* 49(578):54–60.

Scully, Francis J., 1937, Preservation of Plant Material in Natural Colors, *Rhodora* 39(469):16–19.

Sharp, A. J., 1935, Preparing Gymnosperms for the Herbarium, *Rhodora* 37(453): 266–268.

Taylor, T. M. C., 1963, *The Ferns and Fern-allies of British Columbia,* A. Sutton, Victoria, B.C.

Turtox Service Leaflet No. 3, 1959, General Biological Supply House, Chicago.

Wherry, Edgar T., 1949, A Plastic Spray for Coating Herbarium Specimens, *Bartonia* **25**:86.

6

The Protozoans

The phylum Protozoa constitutes a complex of several very old animal groups. These animals are amazingly complex, especially for their size, and are as diversified and well-distributed ecologically as the arthropods and vertebrates. They have invaded most habitats, from marine and fresh-water to terrestrial and parasitic. Whether they should be called one-celled or acellular may remain debatable. Nevertheless, these animals occupy a fascinating taxonomic position and are of interest because of their morphology, physiology, and general behavior. Most techniques for collecting, other than the plankton net, yield relatively few specimens and necessitate some method for concentrating or culturing the catch. Generally, living specimens are exciting for classroom study and thus collecting and culturing methods should be mastered. However, preserved specimens in vials are useful, and with a little practice, excellent permanent or temporary slides may be prepared. With protozoans, it is essential to understand and prepare for adequate field notes, culturing, preservation and storage, and/or permanent slides *before* the collecting is done. Even the transportation of live specimens must be planned ahead of time, as strong light and temperatures other than habitat temperature may kill cultures.

COLLECTING PROTOZOANS

General Remarks

An assortment of clean glass jars with lids, pipettes, fixatives, a plankton net, thermometer, pocketknife, field notebook, and a styrofoam ice chest (to keep specimens cool) are the basic field tools. Take adequate field notes as to microhabitat, season of the year, temperature, and other factors (see Chapter 1), and be sure to label each live or preserved sample with adequate data. As discussed above, preplan field collecting trips so valuable specimens are not allowed to die due to mistreatment. It is best to prepare plankton net hauls, red tides, or other concentrated catches by at least two different fixative systems: one lot in Hollande's fixative (preferably) or Bouin's; the other lot in DaFano's fixative or 4-percent formalin (see Appendix C for formulas and procedure). Alcohol may distort soft-bodied protozoans when used as a killing agent, and also will precipitate sea-water salts, which settle in with the specimens and ruin them. Specimens are stored in 70-percent alcohol, but 4-percent formalin may be used in a pinch for noncalcareous species. Corked vials for storage must be avoided: they allow alcohol to escape and discolor specimens. Rather, place specimens in labeled, screwed-capped vials and store in sealed jars which are also filled with preservative. Glycerin may be added to the preservative to make a 5-percent solution, in order to prevent drying if the preservative evaporates.

Fresh-Water Protozoans

When collecting fresh-water protozoans, provide yourself with a large number of clean bottles with lids. Since culturing is the general rule for these specimens, attempt to collect from as many microhabitats as available. The open surface waters of lakes, ponds, and marshes, permanent or temporary mud puddles, and the like, will be rich with species of flagellate protozoans, some of the actinopod protozoans (such as *Actinosphaerium* and *Lithocolla*), and many species of ciliates. Use a fine plankton net of Number 20 mesh (which contains 173 meshes per inch). Provide the net with a small terminal vial to receive the specimens. Tow the net behind a skiff or cast it from shore and retrieve it by means of a long hand line. In pools too small for active netting, dip up water in a convenient container and concentrate the specimens by pouring the water through the side of a plankton net. For general collecting along shore, a plankton net may be attached directly to a long jointed net handle by various means, as shown in Fig. 6–1. In shallow water containing large quantities of vegetation, dip

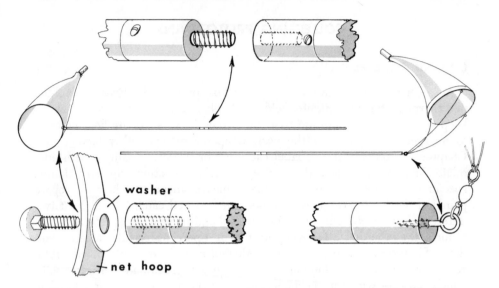

FIG. 6–1. Two methods of attaching a plankton net to a jointed handle. Also see net construction, Chapter 12.

out quantities of greenish looking water, for it may be rich in flagellate protozoans such as *Euglena*. Look for surface scums and collect these by submerging a bottle so that one edge is approximately a millimeter below the surface of the water. The surface layer with the scum will float into the bottle while the bottle is moved along the pond's surface. Such collections may be concentrated through the side of a fine plankton net and then killed or cultured directly.

With a large cooking ladle or tin-can lid, carefully scrape off the upper ⅛ inch of surface mud or sand from pond and lake bottoms and examine this in the laboratory. Soil and mosses adjacent to water hold many protozoans and other organisms that should not be overlooked. Collectors are advised that strong light and temperatures that depart from those of the native habitat will kill cultures. Thus, a water bath, thermos, or insulated foam box may be advisable to receive field collections. A laboratory culture site should be free of high temperatures and bright light also.

Marine Protozoans

Marine protozoans may be found in many of the same situations as those described for fresh-water collecting, above. The uppermost layer of sand or mud from all depths of water is usually populated with living protozoans, including Foraminifera and Radiolaria, and contains the shells of such protozoans which have been deposited on the bottom. Collect samples of sediment (see Chapter 1 for collecting devices) and preserve these as follows: Place the sediment sample in a bottle; permit ample time for the specimens to sink to the bottom if water is present. Next, pour off the superflous water and replace with DaFano's or Hollande's fixative. A solution of 3-percent neutral formalin or buffered formalin (Appendix C) may be used, but this is dangerous, inasmuch as the formalin may become acidic and attack the shells of Foraminifera. By adding a small quantity of rose Bengal stain, eosin, erythrosin, or mercurochrome in a pinch, you will stain the protoplasm of all protozoans that were alive at the time of preservation. Therefore, it will be possible to separate and make a census of the living or recent dead (stained specimens) and empty skeletons (unstained).

The plankton net is also useful for collecting pelagic protozoans, including the Radiolaria and the smaller Foraminifera. This material should be preserved in alcohol after fixation. (See "General Remarks" above.) Large Foraminifera may be obtained by rubbing marine grass and coarse specimens of marine algae over a sieve submerged in a bucket of water. Select a sieve fine enough to trap most of the vegetation, but coarse enough to permit the protozoans to crawl through to the bottom of the bucket. Wash the sieve in the water and concentrate the water by pouring it through the side of a plankon net or through a filter. Kill specimens as described, and preserve in 70-percent alcohol. Some invertebrate filter-feeders and gleaners —such as crabs, clams, and the like—frequently contain large quantities of protozoans and diatoms in their digestive tracts. The gut content of such invertebrates may be preserved directly in alcohol and studied at a later date. When red tides occur, collect the protozoans involved either by means of a plankton net or directly in quart containers. Because of the wide range of species that may occur in plankton blooms, it would be well to have a microscope available to determine the proper method of fixation and preservation for those species on hand. Otherwise, fix portions of the catch with DaFano's and Hollande's fixatives (see Appendix C).

Symbiotic or Parasitic Protozoans

1. The body surfaces and cavities of fresh-water, marine, and terrestrial invertebrates and vertebrates may all harbor parasitic protozoans. Small nodules on the skin or gills of fishes and amphibians often harbor protozoans. Tease such nodules apart on a clean glass slide with a few drops of

isotonic saline. Such nodules may be prepared as temporary slides or smears, or sectioned, using histological techniques.

2. The oral chambers and gill chambers of vertebrate animals frequently harbor amoeboid protozoans. For example, remove some material from around the gums of your own teeth and mix this with saline on a clean slide. Examine this under the microscope with weak illumination for *Entamoeba gingivalis.*

3. The digestive tracts of almost all vertebrates and many invertebrates provide good hunting for protozoans. The lower small intestine and the large intestine, along with the cloaca, harbor many species.

4. Fishes, amphibians, and reptiles may all have characteristic species of ciliates or amoeboid species. Section the gut into small pieces and examine each in turn, both in the general content and along the intestinal wall.

5. Cellulose browsers are excellent sources of protozoans. These can be obtained from termites, roaches, reptiles, and mammals such as rodents, hooved animals, and even elephants according to John Mohr. Protozoans are studied live or prepared for permanent slides. Mammalian gut contents should be collected into a warm thermos. If specimens are to be held a day or more, add four parts saline to one part gut content, and add some suitable oxygen depresser.

6. Nodules on the liver, spleen, and kidneys should all be examined, as well as the content and wall of the urinary bladder.

7. The blood of amphibians, reptiles, and mammals may be rich with haemogregarines or others in the red blood cells, or flagellate protozoans within the lymph. Blood samples should be taken from freshly killed specimens, made into a thin smear, stained, and studied under the microscope. Make a thin smear by putting a fresh drop of blood on the end of a scrupulously clean slide. Place a second slide down into the drop of blood and after a second push this to the far end of the initial slide; stain with Wright's stain (see Appendix C).

SOME CULTURE METHODS

Care of Native Cultures

Field collections may be maintained in the original jars or directly subcultured. Do not overcrowd jars with too much vegetation. Stacking dishes or finger bowls are preferable to jars, in that they provide a high ratio of surface area to volume. Stack finger bowls and cover the top bowl with glass. Be sure that all glassware is meticulously clean and rinsed before collecting or subculturing. Raw or filtered pond water or stream water should be used in preference to tap water which is often contaminated with various metals. Place about ½ inch of pond water in a finger bowl and add some vegetation. Examine cultures under the dissecting microscope for the presence of protozoans. If desired species are abundant, subculture immediately by pipetting them into culture dishes. Usually, however, several species are present in low numbers and require up to two weeks of reproduction before they are numerous. Protozoans show distinct cyclic population effects in culture and will fluctuate periodically from high to low populations.

Physical conditions of the culture dishes must be carefully controlled. Bright light, with occasional sunshine, is required for euglenoid forms; it is tolerable to many ciliates but detrimental to most amoeboid forms which are bottom dwellers. The last-named species are best cultured away from windows, preferably in rooms facing north. Moderate room temperatures are ideal for fresh-water forms, whereas parasitic and marine species require special temperatures. If temperatures tend to approach 90° F., put all culture dishes and jars in a large water bath to reduce temperature fluctuation, since high temperatures are lethal to protozoans. Other critical factors, such as pH and mineral content of the water, may prove important. Native cultures require little or no food until after several weeks have passed. Aquatic protozoans feed on decaying vegetation and bacteria, or are carnivorous and use other species of protozoans for food. Feed as directed below. Finally, many species of protozoans and other invertebrate animals such as rotifers, water fleas, flatworms, and the like, may prove detrimental to cultures by controlling the populations of protozoans. With a fine pipette or filter, remove such organisms. Reculture frequently.

Fresh-Water Amoeboid Protozoans.   Free-living and shell-bearing amoeboid protozoans such as *Amoeba, Difflugia, Pyxidicula,* and *Arcella* are collected among live and rotting vegetation in the bottoms of small lakes, ponds, marshes, etc. When these species are detected in native cultures, transfer them to subculture dishes prepared and kept in subdued lighting at room temperature. A hay infusion has long been used as a food source (Hyman, 1925) but hay seeds, wheat kernels, or rice may also be used. Boil the food source for 5 minutes and add no more than four or five 1-inch pieces of hay per culture dish along with, possibly, a few pieces of grain. Bacteria will be transferred when you are pipetting amoebas into these cultures and will thrive on the vegetable matter, thereby producing a source of food. The decaying vegetable matter will also serve as food, but should be discarded if mold begins to thrive in the cultures. Specimens will become noticeably abundant after two to three weeks and need reculturing after about seven weeks.

Ciliate Protozoans.   Many species of ciliate protozoans in mixed or pure cultures are used in the classroom—most notably *Paramecium*. The food of these ciliates varies from bacteria which grow upon decaying plant or animal matter to living protozoans. *Paramecium* cultures are prepared as described in the previous section, but are kept in moderately bright light (avoid direct sunlight) where bacteria flourish. Transfer specimens by means of a pipette. *Paramecium* populations will reach their peak in from 10 days to 2 weeks (*Turtox Service Leaflet No. 4,* 1959) at which time reculturing is advisable. Predacious ciliates may be kept in *Paramecium* cultures if new supplies of *Paramecium* are added periodically.

Euglenoid Protozoans.   *Euglena* and other similar protozoans can easily be cultured by the Turtox method (*Turtox Service Leaflet No. 4,* 1959). Use clean, wide-mouth, gallon bottles or battery jars filled with boiled pond water or tap water and add a handful of wild grass or hay to each. Age this culture in a sunny place for a week and then inoculate with a *Euglena* culture collected as described above or acquired from a biological supply house. Some additional food (hay or wheat) must be added every few weeks to maintain the culture for a period of several months. Room temperature, with bright light and occasional sunlight, proves best for these protozoans.

Symbiotic Protozoans of Insects.   Common gregarines may be obtained from large mealworms, or by collecting field crickets and keeping these in a dark container with such food as dry dog biscuits, dry vegetation to hide in, and a moist sponge for water. When the digestive tract is dissected upon a slide in saline, *Gregarina* can readily be detected in a large percentage of the specimens. *Trichonympha* and other termite flagellates are best maintained by keeping termites in a dark, covered container. Moist wood, preferably wood that has already been partially worked by the termites, should be used as a food source. Hold a large worker termite by the thorax and gently apply pressure toward the terminal end of the abdomen. This will usually cause the termite to defecate and thus release hundreds of protozoans. Collect the drop of feces on a slide with some saline solution, mix, add a coverslip, and study under the microscope. Frequently these flagellates will go into temporary shock and may remain inactive for a brief period of time. Trager (1934) has worked out two media which may be used for pure cultures of these species.

Literature on Culturing.   Lengthy discussions of culture media for freshwater, marine, and parasitic protozoans are presented by Kudo (1971), Galtsoff *et al.* (1959), Turtox Service Department (*Leaflet No. 4*, 1959), and especially Mackinnon and Hawes (1961).

STUDY METHODS
FOR LIVE PROTOZOANS

Temporary Slides

Wet Mount.   Wet mounts are used for temporary examination of cultures or for classroom study. Clean a slide and coverslip and transfer a few drops of culture material to the slide. Select material either from the bottom (including some bottom debris), from the lighted side of the culture jar, or from the surface of the culture water. Grasp the coverslip between the thumb and first finger and lower it to the slide so that it rests at the edge of the drop of culture media. Finally, support the upper edge of the coverslip with a dissecting needle and slowly lower the coverslip into place. In the process, air will naturally be expelled and few air bubbles will be included in the slide. Examine under the compound microscope. When working with larger protozoans, support the coverslip with cotton fibers or short lengths of hair to prevent crushing.

Fecal Smears.   Fecal material may be obtained from the cages of host animals (frogs, rats, and others), or the intestinal content may be removed during dissection and examination for parasites. Human fecal samples should be collected in clean containers. Here, 1-pint cylindrical containers used for custom-packed ice cream are desirable, in that they are waterproof and disposable. The stool should be firm if the trophozoite stages of amoeboid protozoans are desired. Diarrhetic stools contain predominantly cyst forms. Examine the fecal material within an hour after collecting, as the trophozoite forms quickly die out, losing their taxonomic characteristics.

Place a few drops of saline appropriate to the host animal (Appendix C) on a clean slide. Add to this a small quantity of fecal material and mix this uniformly to make a thin smear. Lower a coverslip as directed above, and seal the coverslip with a soft wax mixture. This mixture is made by adding equal parts of paraffin and Vaseline, melting it, and applying the liquid with a cotton swab. The soft-wax seal is optional, but is recommended when pathogenic species are being examined. Study under the oil-immersion microscope.

Hanging-Drop Technique.   This technique is used to study a limited number of protozoans over long periods of time. Secure a concave slide, a clean coverslip, and a fine pipette. Select a few protozoans from the culture, with the aid of the dissecting microscope, and place these in a small drop of water in the center of the coverslip. Examine the coverslip microscopically to be sure the protozoans you desire are present. Invert the coverslip and carefully lower it into the cavity of the concave slide (Fig. 6–2A). The drop should not be so large as to contact the bottom of the concavity. Finally, seal the edge of the coverslip with Vaseline or soft wax (Appendix C), or preferably silicone culture gum, which, according to Ward's Natural Science Catalog, prevents evaporation and allows CO_2 and O_2 diffusion so so that animals may remain alive for days or weeks.

Use of the Microscope

For most protozoans the standard compound microscope with 50, 100, and 430 magnifications is quite adequate. For the smaller species, especially the parasitic forms, the oil-immersion microscope is used, giving 100, 430, and about 950 magnifications on the three objectives. Controlling the light is very important in locating protozoans. Frequently, light which is too bright will completely obscure protozoans, whereas subdued light enhances their outline. Learn to alter the substage mirror from time to time in order to sharpen the image of specimens under the standard microscope. The use of a substage condenser greatly facilitates study of protozoans and is recommended. Learn to scan a slide under the lowest usable power. This requires that you concentrate and bear in mind the many shapes and forms of protozoans as you scan. When you develop an "eye" for seeing protozoans, low-magnification scanning will save considerable time. Move the slide back and forth under the objective so that you systematically cover the entire area of the coverslip. When you locate a specimen, go to the higher ranges of magnification for study. Guard against overheating or dehydrating specimens. Either seal the coverslip in soft wax or periodically add a drop of water at the edge of the coverslip. Stain may be introduced to slides while they are on the microscope, as discussed below.

Slowing Protozoan Movement

The faster-swimming flagellate and ciliate protozoans move too rapidly for observation of contractile-vacuole movement, digestion, locomotor move-

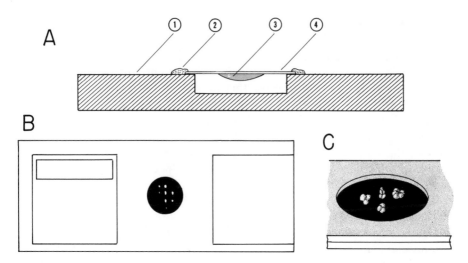

FIG. 6–2. Other slide techniques. A. Hanging-drop technique. (1) Concave slide shown in cross-section, (2) soft wax seal, (3) hanging drop with specimens, (4) coverslip. B–C. Foraminitera-dry-mount slide.

ment, and the like. Several standard methods for slowing protozoan movement are now used, including placing a small amount of cotton or lens paper that has been teased apart on the slide, adding a drop of culture in the center of this material, and supplying a coverslip. The objective is to create small "cells" surrounded by fibers in which the protozoans are contained. Although this method is somewhat satisfactory, the use of methyl cellulose or related alcohol compounds is much more effective. Put some methyl cellulose on a toothpick or dissecting needle and make a ring about ¼ inch in diameter on the slide. Place a drop of culture containing protozoans into this ring and mix the methyl cellulose with the culture water. Finally, add a coverslip so as to exclude air bubbles, and study under the microscope. The very viscous methyl cellulose greatly retards the movement of the protozoans but does not readily interfere with their metabolic activities. One should be most cautious in the use of this compound, however. If the microscope objective becomes coated, clean the lens immediately, for a dried coat of methyl cellulose is very difficult to remove and may damage the finish of the lens.

Vital Stains

Vital stains may be used in dilute concentrations to stain living organisms without interfering with their metabolic activity. Some of these stains are selective for certain structures or materials within the bodies of organisms. Some experimentation may be required to determine the quantity of vital stain necessary. There are three general methods of applying the stain: (1) Add a small quantity of vital stain to a drop of culture medium on a microscope slide, stir with a dissecting needle, add a coverslip, and examine. (2) Add stain after the wet mount has been under examination. Place a drop of stain at one end of the coverslip and "pull" this under the coverslip by applying a piece of paper toweling against the moisture at the other end of the coverslip. With absorption of water from under the coverslip, the stain is forced in and around the specimens. Specimens will show various degrees of staining, depending on the amount of contact they have made with the stain. (3) Add small quantities of stain to the culture until the organisms absorb the desired amount. Then make wet slides in the normal manner. This last technique may be useful for mixing different strains of *Paramecium*, in order to observe conjugation, for example; the different-colored mating strains make the process of clumping and pairing off more obvious.

In addition to staining the cellular content of the organism, vital stains may be used to detect enzyme activity in food vacuoles or to contrast contractile-vacuole activity. For example, feeding *Paramecium* yeast that has been stained with Congo red has been recorded in the literature and is becoming a popular study in laboratory classes. Stiles (1964) suggests mixing 35 milligrams of Congo red with ¼ of a cake of yeast and boiling this in 10 cubic centimeters of water for 10 minutes. When the mixture cools, small quantities are fed to the *Paramecium*. The relative pH can be identified by the color of this vital stain, for pH that ranges from neutral to alkaline registers a brilliant red, whereas pH below 3.0 turns this stain bright blue. Between these levels, purple hues result. Hence, the pH shift in digestion can be observed from the point of food intake to the time of waste discharge. Contractile vacuoles may also be observed by adding fine, particulate carbon to the wet mount. This enhances the appearance of the contractile vacuole and makes it more easily observed. Some of the more common vital stains (which are all mixed with water) are methylene blue, methylene green, aniline yellow, Janus green, neutral red, safranin, and crystal violet. For general laboratory use an assortment of ten vital stains under the trade name "Parstains" is available, along with complete instructions for mixing.

Stain-Kill Methods for Specific Structures

Protozoans and other small organisms are readily stopped and killed by the introduction of strong stains or stains containing alcohol or other poisonous compounds. Alcoholic stains may cause some distortion if they are too strong. Iodine stains may show up starch, cilia or flagella, and nuclei; acetocarmine, methyl green, and crystal violet also contrast the nucleus; Bismarck brown and methylene blue may cause trichocyst discharge from ciliates, and so on. These stains are added to wet mounts that have already been prepared. Pull the stain under the coverslip, as described above, and observe through the low-power microscope during the staining process. Stain, if too dark, may be removed from the culture water by pulling enough fresh water through to dilute the existing stain. Difficulty may arise in pulling or streaming the specimens from under the coverslip. Specimens may be trapped, however, with cotton fibers, or the speed of streaming may be controlled by the rate at which water is absorbed from under the coverslip. For the many additional stains available, consult any good handbook on biological staining techniques (see Appendix B).

PERMANENT SLIDES

For the most part, cytological and histological techniques are beyond the scope of this text. There are many complex methods of staining and counterstaining to show subcellular structures. For animals other than the protozoans such methods are beyond the scope of collecting and preserving. However, in work with the so-called "one-celled" animals familiarity with a few basic techniques of cytological work is essential. Only one method of fixing and one of staining will be presented here. For more complex methods, consult such authors as McClung (1950), Lee (1950), Gurr (1965), and others (Appendix B).

Protozoans are most frequently fixed directly on slides, unless they are available in large quantities. In the latter case they may be concentrated by a slow centrifuge, fixed, stained, dehydrated before being infiltrated with the mounting medium, and transferred to the permanent slide. Smear techniques may be carried out on either the slide or the coverslip. Many media containing protozoans will cause them to adhere to the glass surface. One must avoid drying in all processes of mounting protozoans, in order to eliminate distortion.

Free-Living Protozoans. On the slide (or coverslip) cover an area slightly smaller than the coverslip with Mayer's albumen and allow it to dry until sticky. Next, place a small drop of culture containing flagellates, ciliates, or other protozoans on the albumen and distribute the culture evenly. From this point the author has found the method of Cable (1958), using Schaudinn's fixative and iron-alum hematoxylin, very satisfactory and somewhat less complex than other procedures. This combination of fixative and stain is very satisfactory for all protozoan work. Cable recommends, before smears are made, numbering five slides consecutively with a carborundum or diamond pencil and having ready a 100-cubic-centimeter beaker about two thirds full of Schaudinn's fixative heated to between 60° and 75° C. (Heidenheim's is a good, nonalcoholic substitute fixative.)

Fix immediately by standing the slide on end in a beaker of fixative. Hold the smears apart with pieces of applicator placed across the top of the beaker. Fix for at least 5 minutes. Pass through 30% and 50% alcohol, 2 minutes each. Treat with 70% iodine-alcohol 3 to 5 minutes. Pass through fresh 50% and 30% alcohol 2 minutes each. Wash with two changes of distilled water. Mordant in 2% aqueous iron alum (ferric ammonium sulphate) for 30 minutes or longer. Wash rapidly twice in distilled water. Place in 0.5% aqueous hematoxylin for 30 minutes or longer. Wash twice in distilled water, 1 minute each. Destain as follows: Place all five slides at one time in 1% aqueous iron alum. At the end of one minute, transfer slide number 1 to distilled water, at the end of two minutes, slide number 2, and so on so that the staining time of the various ones will have ranged from 1 to 5 minutes. Change the water at once and let stand 2 minutes. Intensify in saturated aqueous lithium carbonate one minute. Note that the smears become decidedly bluish in color. Dehydrate by passing through 30%, 50%, 70% and 80% alcohol 2 minutes each, 95% 3 minutes, and absolute, 5 minutes. Clear in xylol (3 to 5 minutes). Mount in thin xylo-damar or Permount, using No. 1 coverglasses. Let dry and examine with oil immersion lens and compare staining with the duration of the destaining process.

Surface Scums.   McClung (1950) recommends the following technique for surface scums:

If the culture bears a surface scum in which the Protozoa may be more or less entangled, drop a clean coverglass on the scum, then lift it up with forceps, keeping it horizontal, and a section of the scum will adhere to the undersurface. Tilt the cover and drain off excess water on a piece of filter paper, then drop, scum side down, on the surface of the fixative which has been placed in a convenient container.

From this point Cable's method, as cited above, should be followed.

Fecal and Intestinal Smears.   Material from these sources is self-adherent and needs no albumen. The more or less liquid intestinal content may be placed directly on the slide, whereas harder fecal material should be mixed with a drop or two of saline on the slide. With an applicator stick work the material into a thin uniform smear and transfer it to a fixative, before it has time to dry, as directed above.

Silver Impregnation Technique

Silver stain methods are probably indispensable for ciliary structures, their fibrils and other associated structures, as well as flagellar structures.

The Corliss (1953) modification of the Chatton-Lwoff technique is presented with the author's permission. For more details, consult this reference. For formulas, consult Appendix C.

1. Fix concentrated protozoa for 1–5 minutes in Champy's fluid. Truly hemispherical-bottomed embryological watch-glass, used directly on stage of dissecting microscope, makes a very convenient receptacle for carrying out steps 1–3.

2. Replace Champy's quickly with salinated DaFano's solution, changing twice. Ciliates may be left in this fluid for weeks.

3. Wash out the DaFano's with distilled water.

4. Place small concentrated drop of organisms on very clean grease-free slide; add somewhat smaller drop of warm (35°–40°C) salinated gelatin in solvated condition (powdered gelatin, 10g.; sodium chloride, 0.05 g.; distilled water, 100 ml.). Mix with clean warmed needle. Quickly withdraw excess fluid until specimens remain just nicely embedded in *thin* gelatin layer. It is helpful to reaggregate protozoa in the center of the drop just before step 5.

5. Transfer slide *immediately* from warm stage to cold chamber (covered dish with moist filter paper in bottom, temp. about 5° C). Leave until gelatin has jelled sufficiently: this generally takes 2–4 minutes.

6. Place preparation in cold (5°–10°C) 3% solution of silver nitrate for 10–20 minutes. Keep it cold.

7. Flush slide *thoroughly* with cold distilled water and immediately submerge it, to depth of 3–4 cm., in cold (5°–10°C) distilled water in a white-bottomed dish under a source of ultraviolet light. Good distance of preparation from a 2537 A lamp is 20–30 cm. Change water over slides done in sequence if it becomes warm or cloudy. Expose to light for 10–30 minutes.

8. Remove to cold 70% alcohol; complete the dehydration (85%, 95%, 2 changes 100% alcohol: several minutes in each). Balsam or Permount with number 0 or number 1 coverslip is recommended.

Foraminifera and Radiolaria

These shelled protozoans should be collected, killed in 95-percent isopropyl alcohol, and treated with rose Bengal, as directed above (p. 63). When the specimens have settled to the bottom of the container they may be stored in 70-percent alcohol.

The skeletons of foraminiferans and radiolarians are found in various earth deposits and make up over half of the surface area of the ocean bottom. Sverdrup *et al.* (1949) show that the foraminiferan *Globigerina* makes up a bottom ooze covering over 47 percent of the floors of the Atlantic, Pacific, and Indian oceans, an area of 126.4 million square kilometers. Radiolarian oozes cover about 6.9 million square kilometers. These deposits of tiny skeletons are almost pure in composition and may measure hundreds of feet thick.

Isolating Specimens. Samples of these oozes and other preserved bottom sediments should be processed to remove the mineral content and to isolate the protozoan tests. Very fine mud sediments and coarse sand grains must be eliminated by various methods. One such method is screening through 40-, 80-, and 200-mesh screens to remove the fine muds, followed by the tedious procedure of picking out specimens under the microscope. Secrist (1934) removes clay particles and other extremely minute material by directing a jet of water against the side of a beaker at such an angle as to obtain a whirling motion in the container. This causes the fine material to be momentarily suspended so that it can be removed, leaving the coarser residue at the bottom. This cyclonic movement of water within the beaker also tends to deposit the heavier sand grains in the center; the lighter material, which remains waterborne longer, is distributed toward the outside. Thus, specimens are partially separated from the mineral content. With heavy liquids the differential weight of the mineral content and the animal tests may also be taken advantage of.

Dry-Mounting Foraminiferans.   Special slides are obtainable or may be made for the purpose of dry-mounting foraminiferans. The slide consists of two pieces of laminated cardboard, the upper piece (Fig. 6–2B) is perforated with a round hole about ½ inch in diameter; the lower piece beneath this perforation is painted dull black. If the black surface is not impregnated with a water-soluble glue, a clear, permanent (but soluble) glue must be provided. Foraminiferans are isolated from temporary wet-mount slides without coverslips by pushing the desired specimens to one side with a probe and then picking them up on the moist tip of a fine water-color paintbrush, size 000. The moisture from the brush temporarily activates the glue and fastens the specimens down. When possible, mount several specimens of the same species, each in a different attitude to show all characteristics. The glue quickly adheres to the foraminiferan and will hold permanently or until it is remoistened. The specimen data are recorded on the ends of the slide, and the slides are, in turn, stored in standard slide boxes. Generally, it is not essential to place a clear glass slide over the cardboard preparation. However, this may be useful in very moist climates. For more elaborate mounting systems, consult Schenck and Adams (1943) who detail the operational procedures of commercial micropaleontologic laboratories and present an excellent bibliography.

Other Techniques.   Tests of radiolarians and foraminiferans may be mounted on standard microscope slides with coverslips. Heat-dry samples of foraminiferans, place in xylene for one week, and mount on slides with Permount, using either concave slides, spun cells, plastic rings, or other supports for the coverslip as directed in Appendix B. Dried specimens may be treated for a week in 70-percent alcohol and mounted directly in Turtox

CMC-10. Radiolarians may be mounted directly from water or alcohol in Gray and Wess's mounting medium. Alternately, they may be dried and stored in Euparal Essence for several days and then mounted in Euparal.

REFERENCES

Cable, R. M., 1958, *An Illustrated Laboratory Manual of Parasitology*, Burgess, Minneapolis.

Carson, Carlton M., 1933, Paleontological Notes: A Method of Concentrating Foraminifera, *J. Paleontol.* 7:439.

Corliss, J. O., 1953, Silver Impregnation of Ciliated Protozoa by the Chatton-Lwoff Technique, *Stain Technol.*, 28(2):97–100.

Cushman, J., 1948, *Foraminifera: Their Classification and Economic Use*, Harvard Univ. Press, Cambridge, Mass.

Galigher, Albert, and Eugene N. Kozloff, *Essentials of Practical Microtechnique* (Medical Technology Series), 2nd ed., Lea and Febiger, Philadelphia.

Galtsoff, P. S., *et al.*, 1959, *Culture Methods for Invertebrate Animals*, Dover, New York.

Gurr, Edward, 1965, *Rational Uses of Dyes in Biology and General Staining Methods*, Williams & Wilkins, Baltimore, Md.

Hall, R. P., 1953, *Protozoology*, Prentice-Hall, Englewood Cliffs, N.J.

Hyman, L. H., 1925, Methods of Securing and Cultivating Protozoa, I. General Statements and Methods, *Trans. Amer. Microscop. Soc.* 44:216.

Hyman, L. H., 1931, Methods of Securing and Cultivating Protozoa, II. Paramecium and Other Ciliates, *Trans. Amer. Microscop. Soc.* 50:50.

Jahn, T. L., 1949, *How to Know the Protozoa*, Brown, Dubuque, Iowa.

Kirby, H., 1950, *Materials and Methods in the Study of Protozoa*, Univ. of Calif. Press, Berkeley, Calif.

Kudo, Richard R., 1971, *Protozoology*, C. C Thomas, Springfield, Ill.

Lee, Bolles, 1950, *The Microtomist's Vade-Mecum*, J. B. Gatenby, Ed., McGraw-Hill, Blakiston Division, N.Y.

McClung's Handbook of Microscopial Technique, 1950, Ruth McClung Jones, Ed., Hafner, Darien, Conn.

MacKinnon, Doris L., and R. S. Hawes, 1961, *Introduction to the Study of Protozoa*, Oxford Univ. Press, New York.

Schenck, Hubert C., and Bradford C. Adams, 1943, Operations of Commercial Micropaleontologic Laboratories, *J. Paleontol.* 17(6):554–583.

Secrist, Mark H., 1934, Technique for the Recovery of Paleozoic Arenaceous Foraminifera (Paleontological Notes), *J. Paleontol.* 8:245–246.

Stiles, Karl A., 1964, *Laboratory Exploration in General Biology*, 4th ed., Macmillan, New York.

Sverdrup, H. U., *et al.*, 1949, *The Oceans*, Prentice-Hall, Englewood Cliffs, N.J.

Trager, William, 1934, The Cultivation of a Cellulose-Digesting Flagellate, *Trichomonas termopsidis*, and of Certain Other Termite Protozoa; in a thesis submitted for Ph.D., at Harvard.

Turtox Service Leaflet No. 4, 1959, General Biological Supply House, Chicago.

(See Chapter 1 for other references.)

CHAPTER 7

The Sponges

The sponges, phylum Porifera, are primitive multicellular animals found predominantly in marine waters. They have a cellular grade of construction, since well-developed tissues or organs are lacking. Owing to their primitive development it is difficult to distinguish between individuals and colonies. The name Porifera acknowledges the multitudes of pores, or openings, which allow water to enter the body cavity of the sponge. Collar cells equipped with flagella line the body cavities and are responsible for creating a water current through the body of the sponge and for collecting food.

The taxonomy of sponges may be based on: (1) the gross morphology of the colony or individual (asconoid, leuconoid, syconoid, encrusting, erect, branched, globosed, and so on); (2) the color of the living animal; (3) the substrate—rock, mud, limestone, or mollusk shells; (4) the nature of the skeleton. The skeleton may be made of hard, glass-like calcareous or siliceous spicules or fiber-like spongin. Spicules vary tremendously in relative location, size, sculpturing, and gross shape (see Hyman, 1940). The phylum Porifera is separated into three classes on the basis of the nature and type of skeletal material. The identification of sponges, therefore, depends in part on the analysis of this material.

Collecting Sites and Methods

Fresh-water sponges are found in clear lakes, ponds, and streams in relatively shallow water. They may be collected by hand or with a dredge net (Chapter 12). Remove rocks, heavy pieces of vegetation, or other debris and examine all surfaces for encrusting sponges. Obtain specimens with gemmules which will appear as light-colored bodies scattered throughout the sponge colony. If necessary, collect several times during the year in order to obtain these important reproductive and taxonomic structures. Preserve and dry portions of sponge attached to their substrates, as well as additional material carefully removed with a pocketknife or spatula.

The majority of the marine sponges are found attached to rocks in quiet waters and unprotected and protected rocky habitats, or attached to pilings, floating docks, or other suitable substrata. Most species are to be found in shallow water, although few are found above the zero tide level except on floating substrates. In extremely deep water endemic sponges occur on the softest of muds, supporting themselves by long growths of spicules down into the mud. Collect marine sponges by hand during low tide, placing specimens in wet, nonmucus-secreting algae until the time of preservation. It is essential to keep the specimens cool and moist and to preserve them as soon as possible. Skin-diving will provide many additional specimens. Be sure to turn boulders and look into crevices for sponges growing in highly protected microhabitats. A knife will be essential to remove the sponges from their substrates in most instances.

The biological dredge (Chapter 1) is indispensable when collecting in the intermediate and deep-water habitats.

Killing, Preservation, and Storage

Liquid Preservation.   Sponges may be dried or preserved in liquid. For liquid preservation, wash specimens in water to remove debris and preserve in 75- to 95-percent alcohol. The weaker solution is suitable for small specimens, but large specimens or specimens crowded in a single bottle should be preserved in 95-percent alcohol. Change the alcohol after 24 hours on large or crowded specimens. Do not use formalin for preservation, as it will destroy calcareous spicules and break down the sponge "tissue." Neutralize formalin if it must be used and transfer specimens as quickly as possible into strong alcohol. Specimens are stored in the museum in strong alcohol. Small specimens should be kept in alcohol-filled, cotton-stoppered vials, along with their field data. These vials, in turn, are placed in a larger museum jar filled with alcohol. Large specimens are kept in single jars, with the field data included.

Dry Preservation.   Dried sponges are very suitable for taxonomic purposes and cost less to prepare, but are not too useful for beginning biology students. If sponges are dried, additional small individuals or portions should be preserved in alcohol. Sponges may be dried, with or without their substrate, by being washed and placed in a warm place out of the sun. Before drying large specimens, attach labels by means of a string threaded through the body of the sponge. When they are thoroughly dried, sponges should be kept in small cardboard boxes and supported with tissue paper. Field data and taxonomic data should be included with each specimen. Paradichlorobenzene or naphthalene flakes may be added to dry containers, though this is usually not necessary.

Fixation for Histology

Sponges intended for "tissue" studies must be fixed in the field with Champy's fixative or other standard fixatives containing osmic acid. Wash specimens to free them of debris and place them directly in the fixative for about 6 hours. Subject the fixed specimens to running water for an additional 6 hours and transfer them to strong alcohol. Do not leave specimens in the fixative, as any water-bearing solution will break down the tissues. On returning to the laboratory, remove the specimen's spicules, stain, and section as described in handbooks of microscopic technique (Gatenby and Beams, 1950, and others).

Skeletal Material

Sponge skeletal material consists of fibrous spongin or of siliceous or calcareous spicules. Megascleres are larger spicules which make up the loose inner mass of support; microscleres are peripheral and make up small dermal spicules. Spicules must be studied independently and occasionally *in situ* to show the spicule arrangement.

Spongin.   Sponges containing spongin should be washed in fresh water and dried. Beat the sponge to loosen the cells mechanically, rewash, stain with safranin or eosin, and dry. Remove small portions and mount in Permount or balsam slides.

Siliceous Spicules.   Select a sample piece of sponge and remove the soft tissue by boiling it in several drops of concentrated nitric acid on a slide over a flame. When it is dry, add a mounting medium (Permount, balsam, or either) and a coverslip, and store in a horizontal position until dry. A better preparation with less debris is obtained by treating sponge overnight in a test tube containing 6 or 7 cubic centimeters of nitric acid or Clorox. In the morning, carefully pour off the acid, leaving the spicules at the bottom, and wash with several changes of water. Permit the spicules to settle to the bottom after each washing. Exchange the water for alcohol and transfer a drop of the alcohol–spicule mixture to a slide. When completely dry, add mounting medium and a coverslip.

Calcareous Spicules. The composition of spicules in unknown sponges may be determined by treating a small portion of the sponge in acid. Calcareous spicules will react with the acid: their "boiling" testifies to their composition. Treat calcareous spicules as described under siliceous spicules, but use a 10-percent potassium hydroxide (KOH) solution in place of the nitric acid. After the treatment, wash, transfer to alcohol, and prepare the slides, as described above.

Spicule Arrangement. To determine the spicule arrangement *in situ*, remove a small but undisturbed sample of sponge and dry. When it is completely dry, wet with xylene, transfer to a xylene–balsam mixture, and then transfer to a slide. Mount in balsam or Permount, using broken pieces of a slide or coverslip to prop up the cover glass if necessary.

REFERENCES

Hyman, Libbie H., 1940, *The Invertebrates: Protozoa through Ctenophora*, McGraw-Hill, New York.

Jewell, M., 1959, "Porifera," in *Ward and Whipple's Freshwater Biology*, W. T. Edmondson, Ed., Wiley, New York, pp. 298–312.

Lee, Bolles, 1950, *The Microtomist's Vade-Mecum*, J. B. Gatenby, Ed., McGraw-Hill, Blakiston Division, New York.

Needham, James G., et al., 1937, *Culture Methods for Invertebrate Animals*, Dover, New York.

(See Chapter 1 for other references.)

CHAPTER 8

The Coelenterates

Fresh-Water Hydra

Collecting. The general habitat of the hydra is the small fresh-water lake or pond and small moving streams and rivers, especially where they flow from lakes. In particular, shallow waters about one foot in depth produce the best light and temperature combination for hydra. Specimens may be seasonal where the winters are severe, but are generally available through the late spring, summer, and autumn.

Evicting Hydra. Hydra may be dislodged from their substrate by squirting them with a jet of water from a pipette. To transfer them to a new culture, pick them up quickly with a pipette which has a moderately large opening. Be sure to expel them with equal rapidity; otherwise, they will cling to the inside of the pipette. When hundreds or thousands of hydra cling to rocks or vegetation, bring this material into the laboratory, place it in a fairly dark area, and allow it to stand several hours or overnight. The hydra will dislodge themselves and float to the surface as the available oxygen decreases in the container. They can then easily be removed with a small aquarium net or pipette and transferred to a new culture dish.

Fixation and Preservation.   Transfer large numbers of hydra to Petri dishes with about 3 millimeters of water. After allowing them to become fully expanded, pour hot Bouin's fixative (50° C.) over them. After a few moments pour off the diluted fixative and add fresh Bouin's fixative for 30 minutes. Following this, wash in several changes of 30-percent alcohol, move the specimens through 50-percent alcohol, and store in 70-percent alcohol. In this procedure, be prepared to fix the specimens as soon as they become fully expanded; prolonged waiting may cause them to partially contract and remain that way. When specimens are few, treat them individually so that each may be caught at the peak of its expansion.

Preserved hydra should be placed in vials of alcohol, along with their collecting data, and kept in a large airtight jar of 70-percent alcohol. These specimens may be used on temporary slides, stained with methylene blue, neutral red, or other standard laboratory stains, and provided with supports for the coverslip (Appendix B). For permanent slides, stain with borax carmine or alum hematoxylin, dehydrate through absolute alcohol, transfer to xylene, and mount in balsam or Permount (see Appendix B).

Collecting and Seasonal Studies.   Hydroids, being filter feeders that require large quantities of microscopic organisms, live in the upper layers of ocean water and are generally attached to rocks, pilings, and other suitable substrates. They may be found offshore by pulling in giant brown kelp and examining the fronds and holdfasts. It will be noted that hydroids will grow in great profusion on certain levels of the pilings and corresponding levels of rock substrates. Another good method of collecting is to trap hydroids on wood blocks (2 by 2 by 6 inches) placed at 1- or 2-foot intervals on a weighted rope, suspended from a floating dock for a month. These blocks remain at a constant depth relative to the surface and thus attract most phyla of sessile invertebrates. A small hand dredge (see Chapter 1) will provide rocks and other objects used for attachment by hydroids. Also, visit ocean beaches immediately following severe storms when seaweeds, and even rocks to which they cling, come to shore bearing a wide array of invertebrates. Use a pocketknife to free hydroids. Keep specimens in cool sea water; narcotize as soon as possible.

Narcotizing, Killing, and Preserving.   Place small colonies in jars of clean sea water, in a cool semidark area. When the polyps are fully expanded add small quantities of epsom salts (see Appendix D). Continue to add small quantities every 20 to 30 minutes for several hours, being careful not to disturb the polyps. After 3 or 4 hours probe one or two polyps with a dissecting needle and when they prove insensible to touch, carefully add enough formalin to produce a 5-percent solution. If narcotizing compounds are not available, kill specimens with formalin and maintain in 5-percent neutral formalin or 70-percent alcohol after going through intermediate changes of 30 percent and 50 percent. Place specimens in a beaker of sea water and, when they are fully expanded, siphon off all water except enough to cover the specimens without disturbing them. With great caution, pour in a solution of corrosive sublimate (heated to 50° to 60° C.), then cool the solution, wash in several changes of fresh water, and preserve as above. (Consult Appendix C for the use and neutralization of corrosive sublimate.) Good results may also be obtained by following Bianco's method, using hot Bouin's fixative (50° to 60° C.) as the killing agent. If slides are desired, stain with borax carmine, dehydrate, and mount on slides in balsam or Permount with coverslip supports as directed in Appendix B.

Medusae

The small medusae of the Hydrozoa and the jellyfish forms of Scyphozoa will be treated simultaneously. A number of methods are available for specific groups of medusae, but as Bianco (1899) states, different species within a single genus may require different techniques. Obviously, some experimentation is required in testing the different methods listed. For general use, most of the medusae may be preserved directly in sea water with 5-percent neutral formalin.

Collecting. Medusae are pelagic forms found in most strata of water from the surface down to the aphotic, or lightless, zone. Bays, harbors, and estuaries tend to swarm with medusae during the summer months. Because they are extremely soft, they can easily be destroyed in general collecting nets such as tow nets, plankton nets, bathy-pelagic nets, and the like. Whenever possible, it is best to dip specimens out directly with a net or a bucket. Regardless of the means of collection, however, transfer specimens immediately to containers of sea water. The best-preserved specimens are those which are narcotized and preserved while in a living condition. Otherwise, add several drops of formalin to kill. See next paragraph for storage.

Various Narcotizing and Preserving Techniques. The small medusae from hydroid colonies, such as *Obelia,* or small scyphozoan jellyfishes, may best be narcotized with ethyl urethane (Galigher, 1934) or with Chloretone or epsom salts. Place specimens in a shallow dish of sea water, add a few crystals of urethane or Chloretone, and let stand for several hours. With epsom salts, add crystals periodically, as described under colonial hydroids. When the medusae are insensitive to probing, pipette out most of the water and kill them with 5-percent formalin. Store specimens in this solution or run them up through the alcohols to 70 percent.

For members of the Siphonophora (*Velella, Physalia,* and others), or larger scyphozoans, fairly good specimens may be obtained by direct preservation in 5-percent formalin. Store specimens in 5-percent formalin or 70-percent alcohol. Always consider the fluid content of the animal when measuring the preservative. For better specimens, narcotize with menthol crystals (Appendix D) or epsom salts, as directed above.

Anemones

Collecting.   Anemones are either sessile, attached organisms or sand-dwelling burrowers. They are capable of some locomotion, creeping along on the pedal disc or even swimming. Sand-dwelling species are simply dug from the substrate; attached forms may be collected along with part of their substrate or by slipping the blade of a very dull knife under the pedal disc. The specimens should be kept in clean sea water and should not be overcrowded with other organisms. Deep-water species are obtained with the biological dredge (Chapter 1).

Narcotizing and Preserving.   Sea anemones must be expanded in sea water, and then narcotized, before they may be preserved. The author has had outstanding success with many species of sea anemones, using clove oil and Chloretone. Regardless of the method, however, it is best to treat each anemone in a separate jar of sea water, and to begin narcotizing in the evening so that specimens may stand all night under the influence of the particular drug. Several large drops of clove oil were added to each quart jar (¾ full), once each hour for 3 hours, beginning at 8 o'clock in the evening. The amount of clove oil was increased somewhat during the last addition. By 8 o'clock the next morning specimens should be well expanded (provided they were expanded before the treatment began) and completely insensitive to formalin. Probe the tentacles lightly to determine if the animal is sensitive to touch. If it is not, add enough formalin to make a 5-percent solution. The Chloretone method is essentially the same as the clove oil method. Add a moderate pinch of Chloretone crystals each hour for 3 or 4 hours, in slightly increasing amounts, during the evening. Permit the specimens to stand overnight, test them for sensitivity by probing the tentacles, and preserve in formalin when they become insensitive.

Epsom salts have long been used to anesthetize sea anemones. The writer finds that this drug is less effective if it is used as directed for Chloretone above; thus, epsom salts should be added in small amounts every *half hour,* for a period up to 24 hours, until the animals are insensitive to touch. One problem here is that animals tested prematurely may contract the entire body and fail to reexpand under the effects of epsom salts. Therefore, gently tease one or two tentacles, rather than probing the entire animal. When the specimen is totally insensible, preserve it in 5-percent formalin. Gohar (1937) found that some anemones began to macerate before they were completely narcotized. Thus, he suggests beginning the epsom salts treatment with specimens placed in 50 to 100 times their own volume of sea water. During this process add three drops of 1-percent formalin for every 100 cubic centimeters of water, every 15 minutes, and double the volume added each hour until the specimens are insensible; then preserve in 5-percent formalin.

Another technique that will work for sea anemones is to let expanded specimens stand in containers of water at room temperature, from 24 hours to several days. Such specimens will gradually become insensitive as the oxygen is used up and as the water becomes polluted with excreta. This process may be hastened by beginning with chilled sea water which was previously boiled to remove the free oxygen. A second method of hastening the process is to add some organic material to further pollute the water and deplete the oxygen.

Some sea anemones intended for histological work should be expanded in beakers of sea water. When the specimens are fully expanded, siphon off most of the water and kill them by pouring in hot Bouin's fixative (50° to 60° C.), hot 20-percent formalin, or hot FAA.

Horny Corals, Sea Pens, Sea Whips, and Others

Collecting.   The horny corals grow in shallow to moderately deep water, attached to reefs or isolated stony substrates. Deep-water forms are collected by dredging; shallow species are obtained by skin diving. A geology pick is quite suitable for cracking off chips of rock which support the horny corals. They may also be dislodged by forcing a heavy knife between the holdfast and the substrate. Sea pens, sea pansies, and the like, may be obtained at low tide or by dredging. Be sure to take field and color notes as described in Chapter 1.

Narcotizing and Preserving.   The horny corals are very useful when dried. Dip specimens in neutral formalin for 15 minutes and dry them in a warm but shaded place. Prior to drying, arrange the branches so that they will take up the least amount of space in a museum tray, if this is important. Always secure a waterproof tag containing the field data to the stem of each specimen before drying. After the specimens are thoroughly dried keep them in light-proof boxes with a few crystals of paradichlorobenzene. Samples of horny corals and all sea pens and related species should be narcotized and preserved. Permit these specimens to expand normally in containers of sea water and narcotize as described above under "Anemones," or kill them quickly with hot Bouin's, hot corrosive sublimate, or hot formalin.

Stony Corals

Collecting. Stony corals are dominant forms in tropical marine waters, occupying the intertidal zone and extending down to considerable depths. At the higher latitudes the number of species decreases markedly, until only one genus, *Balanophyllia,* a solitary coral, is found intertidally and a few other species are found in deeper water extending on to the north. When a rare coral is collected, part of the rock substrate to which it is attached should be chipped away, along with the specimen, by means of a cold chisel and hammer. Where corals abound they may be collected while you are skin diving, with a geology pick or by means of a dredge.

Narcotizing and Preserving. When expanded specimens are required, narcotize and fix as described under "Horny Corals." Use 70-percent alcohol for a preservative or 5-percent neutral formalin. Straight formalin will greatly damage the skeleton of any coral.

Dried coral specimens are the most useful in spite of the fact that they are fragile. There are three general methods for preparing dried coral; the second method is perhaps the best. To prepare corals by the first method, simply wash freshly collected coral to remove mucus and foreign debris. Next, place them in a well-ventilated area to dry, but avoid constant sunlight. When the specimens are thoroughly dry, store in cardboard containers along with their field and taxonomic data. This technique is especially useful for those species with very small polyps (feeding animals).

The second and most useful method for preparing dried coral involves the removal of all tissue prior to drying. Place specimens in containers of water for three to four days. Following this, wash the specimens under a hose to remove rotted tissue and debris. Next, place the specimens in a 10-percent solution of household bleach for one day, or until the specimens are whitened, rinse them thoroughly in several changes of fresh water, and then permit them to dry.

The last method calls for removing tissue by soaking specimens in a 10- to 20-percent solution of household bleach from one to three days, hosing thoroughly, and rinsing in several changes of water before drying. This method may yellow some specimens. KOH or NaOH are preferred to bleach as they do not distort the delicate septa microscopically.

REFERENCES

Bianco, S. L., 1899, The Methods Employed at the Naples Zoological Station for the Preservation of Marine Animals, *Bull. U.S. Nat. Mus.* 39(M):1–37.

Galigher, A. E., 1934, "The Essentials of Practical Microtechnique," published privately.

Gohar, H. A. F., 1937, The Preservation of Contractile Marine Animals in an Expanded Condition, *J. Mar. Biol. Assoc.* 22(1):295–299.

Pennak, Robert W., 1953, *Fresh-Water Invertebrates of the United States,* Ronald, New York.

Smith, F. G. W., 1948, *Atlantic Reef Corals,* Univ. of Miami Press, Coral Gables, Fla.

(See Chapter 1 for other references.)

CHAPTER 9

The Lower
"Worm" Phyla

PHYLUM PLATYHELMINTHES

Class Turbellaria

Collecting Fresh-Water Species. Collect fresh-water turbellarians in lakes, ponds, marshes, streams, and eddies or back runs of larger rivers where the water is a foot or less in depth. These animals are photonegative and thus are inactive during the day and are found in places of hiding. The undersides of rocks (except those firmly buried in mud), submerged logs or boards, lily pads, and the like may harbor specimens. Algal turfs and scums on rocks are frequented by smaller specimens. Standing pond vegetation should also be examined. One method of obtaining small individuals is to collect large quantities of pond or stream vegetation into buckets and store them in a dimly lighted area. As the oxygen tension changes turbellarians will be seen on the sides of the bucket or on the surface of the water. Replacing a large quantity of the water with freshly boiled and cooled water may hasten the process.

Shallow streams may be baited with pieces of raw beef or liver by placing the meat between rocks or other debris where turbellarians may occur. The bait should be checked every half hour or, if it is left overnight, the hiding places in the vicinity of the bait should be examined for specimens. A simple trap can be fashioned: put a meat bait in a small-mouth bottle (aspirin, or other), tie it to a string, cast it from shore, and retrieve it periodically for examination.

Collecting Marine Specimens.   Many of the methods mentioned above are adequate for marine collecting. The prime requisite of the larger triclads and polyclads is a suitable hiding place. Thus, crevices where one rock covers another, places where rocks cover coarse sand or gravel substrates, clusters of barnacles and mussels clinging to rocks or pilings, holdfasts of algae, discarded clam and snail shells, gill chambers of some arthropod animals, and the like, may all harbor these animals.

Narcotizing, Fixing, and Preserving.   Turbellarians are either prepared on slide mounts or preserved in 70-percent alcohol or 5-percent formalin. In either case, specimens must be well expanded before they are fixed and preserved. This presents little difficulty with the small individuals, but may require narcotizing or other methods for larger specimens. FAA (Appendix C) is one of the best all-around fixatives, and may also be used for temporary storage. Gilson's fixative (Appendix C) is also widely used. However, specimens should be kept in Gilson's fixative for 24 hours or longer to ensure thorough fixation. A 5-percent formalin solution (mixed with sea water for marine specimens) is quite suitable for general preservation where specimens are not intended for slide making.

Fix specimens as follows: transfer small specimens in a pipette with a minimum amount of water to the bottom of a dry Petri dish or a glass plate. The water should be just sufficient to permit the animal to expand. When it is thoroughly expanded, squirt on hot (50° to 60° C.) fixative with a pipette. After several minutes pick up the specimen with a thin spatula and transfer to a vial of fixative for 3 or 4 hours (FAA) to 24 hours (Gilson's fixative). Store in 5-percent formalin or move specimens up through the alcohols to 70 percent.

Moderate-sized turbellarians should be transferred to a glass plate with a small amount of water. When they are fully expanded, drop a coverslip on top of the specimens. The more muscular specimens may be covered with part of a microscope slide. The volume of water should be small enough so as not to float the slide or coverslip, but rather to create suction which holds the worm in an expanded position. Add FAA to one end of the coverslip and pull this under and around the specimen by absorbing water, with a piece of paper toweling, at the opposite end. Again, be careful not to introduce enough liquid to float the coverslip and allow the animal to contract. Specimens should be treated up to ½ hour with additional quantities of FAA, added when necessary. Finally, lift the coverslip and slide the specimen into a container of fixative. Treat as above. An older method for dealing with very large turbellarians was to place them between two microscope slides which were subsequently bound with rubber bands. This preparation was then placed in a container with fixative for 2 to 5 hours, depending on specimen size, and then transferred to a proper fixative and treated as above.

Narcotizing probably gives the best results with large specimens. Place specimens in a small dish with water from their normal habitat. Add crystals of Chloretone, chloral hydrate, or menthol in small amounts for up to 3 or 4 hours. When the animals are insensitive, place the specimen on a glass plate, cover with a glass slide, and fix as directed above.

Lavoie (1958) found the following method very suitable for large polyclad turbellarians: Transfer specimens to a glass Petri dish with no excess water, and place in the freezing compartment of a refrigerator for about 20 minutes. Remove from the freezer and immediately fix with an FAA solution, prepared as follows: 90 parts of alcohol (95 percent), 5 parts of formalin. Heat this solution and pour directly over the specimen. Lavoie reports little shrinking or loss of natural color, and no histological damage with this method.

Slide Preparation.   Lightly pigmented specimens may be cleared and mounted in Permount or balsam, or lightly stained and mounted. Borax carmine (described below) is a good general stain. Pennak (1953) recommends orange G and Delafield's hematoxylin for whole mounts of rhabdocoels (and probably other forms). Darkly pigmented specimens should be bleached in undiluted hydrogen peroxide, after which they are run up through the alcohols to 70 percent, stained, and mounted.

Galigher (1934) suggests feeding planarians on liver pulp ground with carmine prior to fixation with 10-percent formalin. Specimens are then dehydrated, cleared, and mounted unstained.

Borax-Carmine Stain.   Transfer specimens from 70-percent alcohol into borax carmine and let stand until the specimens are thoroughly permeated, sometimes one or more days. Following this, destain in 70-percent acetic alcohol (Appendix C) until a bright, translucent pink appearance is obtained. Use the dissecting microscope every few hours, for as long as two days, to determine the progress of destaining. Finally, wash in neutral 70-

percent alcohol, dehydrate in successive washes of alcohol through 95-per-cent and absolute alcohol. Next, clear in neutral beechwood creosote, clove oil, or xylol and mount in Permount or balsam with supported coverslips. An alternative method is to mount directly from the 70-percent neutral alcohol in Turtox CMC-10 mounting medium. This medium permits objects to be transferred directly from water or alcohol; it will both mount and clear specimens, but should be sealed with asphaltum upon drying.

Turtox CMC-S is a stain-mounting medium designed for small inverte-brates and is excellent for fixed flatworms. Living specimens may be killed, stained, and mounted when placed in this medium on a slide with a cover-slip. Tissues with similar refractive indexes sometimes become uniformly colored, since this medium has no destaining process. This process can be overcome to some extent in smaller specimens by mixing 1 part of CMC-S with 1 part of CMC-10 before using. Ring with asphaltum when dry.

Class Trematoda

Collecting Adults. Trematodes are parasitic animals which are non-segmented, lack cilia, possess a cuticle in place of the outer epidermis of the turbellarians, and possess suckers for attachment. They are most com-mon as parasites among vertebrate animals, especially those associated with water. Their life cycles are quite complex, a multiplicity of larval forms frequently involving intermediate hosts. Trematodes may be found on their hosts in nasal chambers, gills and gill chambers, in stomachs (rarely), commonly in the small and large intestines, cloaca, lungs, and urinary bladder, or hiding under scales of fishes. Gills of fishes and amphibians should be placed in a saline solution (Appendix C) with some Chloretone crystals, set aside, and examined later under the dissecting microscope. Trematodes hiding there become relaxed and may be "combed" out of the gills with a dissecting needle. They will be found by carefully examining the gills and the bottom of the dish.

The complete digestive tract of the host should be examined. Remove the entire tract and place it in a dish of saline, then cut it into short sections, split each section, and examine it for trematodes. Large trematodes will be quite obvious, whereas small ones will hide among the villi of the intestine and must be located with the dissecting microscope. When adult specimens are located, remove them by dislodging the suckers; the finger-nail or a dull scalpel may be forced against the worm in such a way as to dislodge the sucker without injuring the worm. Transfer worms to dishes of saline to clean them prior to fixation.

Cut lungs of possible host animals open in saline and examine under the dissecting microscope. Likewise, puncture the urinary bladder, draining it of its contents, cut open, and examine in a dish of saline. Also examine the urine drained from the bladder. Finally, look for nodules under the scales of fishes, and elsewhere, which may harbor adult or larval trematodes.

Fixing and Narcotizing.   Trematodes are almost always prepared for microscopic slides in that the internal anatomy is of paramount taxonomic importance, whereas external anatomy is of little value except for general display of large specimens. Therefore, all adults are fixed while under the pressure of a coverslip or slide, with or without narcotizing, depending on their size. Follow all of those techniques presented for the Turbellaria when working with the Trematoda.

Slide Preparation.   Borax carmine will yield excellent results when used with the Trematoda. A mixture of Turtox CMC-S and CMC-10 will also give fair results. Follow the instructions under "Turbellaria." Hematein, described for tapeworms below, is also very useful.

Collecting Larvae.   Monogenetic trematodes have more or less direct life histories, whereas the digenetic trematodes may have as many as six distinct larval forms. Some of the larvae, in turn, are parasitic and may infect one or more hosts during the life cycle. The miracidium hatches from an egg released in water, and locates an intermediate host, usually a snail. After penetrating the snail, it metamorphoses into a sporocyst stage which may, in turn, produce either the daughter sporocysts or the redia stage. Either sporocyst or redial stage gives rise, ultimately, to the cercaria stage.

Place adult worms removed from the final host in a dish containing an appropriate saline (Appendix C). When eggs are expelled, transfer them to the water used by the final host (pond water or sea water). Observe the water every few hours for miracidia by placing the culture dish over a dark surface and shining a light through from the side. Swimming miracidia may be pipetted into formalin or FAA and stained and mounted as adult specimens; the Turtox medium is useful here.

Snails can be treated in one of two ways. Place them in finger bowls with the habitat water, and observe daily for cercaria being expelled. This is the best way to get mature, characteristically formed cercaria. The other method for obtaining cercaria, and the only means of getting sporocysts or redia, is to dissect the snail. Gently crack the shell without crushing the soft parts of the snail. Pick away the shell or unwind the animal from the shell case.

Those coiled portions occupying the tip of the shell (the gonad and liver) are the best hunting grounds. Place the evicted snail in a dish of saline and gently tease apart the gonad and liver. Many animals will be so infected that the gonad will consist of a membranous sac crawling with sporocysts, redia, and cercaria. Look through the dissected debris and the water on the bottom of the dissecting microscope. Remove larval forms to a clean dish of saline with a small pipette.

Preservation of Larvae.   Larvae are stained and prepared in the same manner as adults. Place larvae in small quantities of water on a glass plate and fix with hot FAA, without coverslips, as described above. Transfer specimens to a vial for staining. Permit these specimens to settle to the bottom after each successive wash, or gently spin them to the bottom with a hand centrifuge. Mount in balsam or Permount.

Class Cestoidea

Collecting True Tapeworms.   The subclass Eucestoda is comprised of segmented worms possessing a scolex for attachment and larvae bearing only six hooks. There are a number of different larval forms, depending on the order of tapeworms involved. Tapeworms occupy the digestive tract, most frequently the small and large intestines, of all of the vertebrate animals. Moreover, both herbivores and carnivores become infected with tapeworms. Most tapeworms shed mature segments and never become extremely large, measuring up to 30 inches or so; others, which do not shed their segments, may grow from 10 to 20 feet long. Eviscerate the host animal and carefully open the digestive tract, looking continuously for chains of tapeworm segments. If no large tapes are found, cut the tract into shorter sections and submerge in saline. Smaller tapes may appear as very thin whitish threads a few inches in length. When you locate a tapeworm, do not extract it, but carefully locate the scolex (at point of attachment). Observe this under the dissecting scope, and with a dull scalpel blade or probe attempt to dislodge the hooks or suckers. Tapeworm bodies are seldom strong enough to support even their own weight and will break very readily. The scolex is essential for identification and must be secured.

Larval tapeworms may be lodged in almost any organ of the body. They are common in the liver, spleen, lining of the body cavity, and in the fat layers under the skin, especially of marine mammals. Larval tapes may appear sac-like, pigmented or white, fibrous, or cyst-like, and are, in general, difficult to recognize. Consult Wardle and McLeod (1968) or textbooks on parasitology for the appearance of larval tapes.

Preservation of Tapeworms.   Since internal structures are of paramount taxonomic importance, adult and larval tapeworms are generally prepared on slides. Both, however, may be fixed and stored in 5-percent formalin or even displayed against black glass as described in Appendix A. Transfer small tapeworms to a glass plate with a large open-mouth pipette. Arrange these specimens so they will conveniently fit under a coverslip, blot away excess water, add a coverslip to hold them in place, and fix with FAA. Place the FAA at one end of the coverslip and pull through by absorbing the water at the other end with a piece of paper toweling. Stain with borax carmine, as described under "Turbellaria." Large specimens are cut into sections small enough to go under coverslips. The "head" (scolex), "neck," and first few segments are kept intact, some immature segments are preserved, some mature segments from farther down the chain are preserved, and some of the terminal segments which should be ripe are also secured. After cutting them into convenient lengths, place specimens on a glass plate, cover with coverslips or slides, and fix with FAA as described above. More "muscular" and robust tapeworms have to be relaxed in their entirety in order to render them flat enough and thin enough for taxonomic uses. Place the entire animal in a dish of saline and put this in the refrigerator (not the freezer) until it is thoroughly chilled. Next, lift the specimen to a glass plate which also has been chilled. The plate should be long enough to accommodate the entire tapeworm or sections of at least 16 inches. Arrange the specimen in a straight line and blot the water away from the anterior end. Next, gently hold the anterior end and gently pull the opposite end out on the glass plate, in order to stretch the specimen. Remove excess water from the distal end so that it will not contract again, cover the desired sections of the tapeworm with glass slides, and fix by adding FAA for 30 minutes. Specimens may be stained in borax carmine (see p. 90) or with hematein.

Staining with Hematein.   Dehydrate the specimen by treating with 50-percent and 30-percent alcohol for 30 minutes or longer with each. Wash in distilled water, add stain (Appendix C), and let stand for one hour. Following this, wash in two changes of distilled water for 10 minutes each. Destain in 30-percent alcohol until the color changes from blue to medium pink. Next, treat in ammonia water for about 20 minutes to intensify the color. Wash rapidly in three changes of distilled water, move through 30-, 50-, 70-, 85-, and 95-percent alcohol for 25 minutes each (or longer with large specimens), wash in two additional changes of 95-percent alcohol and move to absolute alcohol, clear, and mount (see Appendix B for further instructions on slide-making techniques).

PHYLUM ASCHELMINTHES

Class Rotifera

Collecting Rotifers. Rotifers are aquatic animals of two general types: the pelagic (free-swimming) and the benthic (bottom-dwelling). Lakes, ponds, marshes, brackish water, and the like should be sampled. Clear water and stagnant water (which is usually fertile and highly congested with microorganisms) should be sampled. Free-swimming specimens are obtained with the standard plankton net (125 meshes) or the coarse plankton net (75 meshes) which is towed through the water (Chapter 1). Specimens should periodically be transferred from the collecting vial attached to the tip of the plankton net to a clean bottle of water. Bottom-dwelling specimens are best collected along with plant debris. Sample different kinds of plants from various depths, including rotting vegetation, and keep separated in clean jars two-thirds full of water.

Place the jars with plant debris near a north window in the laboratory. Push the plant debris toward the side of the jar away from the light. Rotifers will migrate toward the light and can be pipetted into a vial.

De Beauchamp (1912) recommends collecting both wet and dried mosses and scrapings from mud surfaces of dried pools. These are placed in culture dishes and supplied with pond water which was previously boiled, cooled, and shaken well in a jar to add oxygen. Eventually, rotifers will emerge from dormant stages or from eggs contained in the moss or mud surfaces. These are then transferred to a vial for narcotizing. Rotifers vary in abundance seasonally throughout most of the United States, having two population peaks which probably correspond with nutrient cycles: April–May and September–October.

Narcotizing and Fixing. Harring and Myers (1922) recommend narcotizing specimens placed in a watch glass which can then be observed under the dissecting microscope. Add a few crystals of Chloretone (in place of coccine hydrochlorate, which is no longer available) to the specimen and observe it until the cilia almost stop moving. Quickly add a few drops of 1-percent osmic acid directly on the animals and mix well. Wash through two or three changes of water, from 10 to 15 minutes each, to remove the osmic acid. If the specimens are darkened, bleach in a weak solution of potassium hydroxide, and rewash. Store in 2- to 5-percent formalin, or run up through 30- and 50-percent alcohol and store in 70-percent alcohol.

Other compounds have given good results in relaxing specimens. A freshly made, 1-percent solution of hydroxylamine hydrochloride or a 2-percent solution of strychnine sulfate (*Turtox Service Leaflet No. 2*, 1959), 2-percent benzamine lactate, 2-percent butyn or 2-percent hydroxylamine hydrochloride (Pennak, 1953), or 1-percent neosynephrin hydrochloride (Myers, 1942) have been used.

Temporary Slide Mounts.   Harring and Myers (1922) suggest the following: Secure depression slides with the concave measuring about 10 millimeters in diameter by ½ millimeter in depth, and place a drop of liquid containing preserved specimens on the inside edge of the depression. Lower a large, square coverslip down over the drop, and then push the coverslip so it covers the concavity. Place the slide on a microscope and tilt the microscope back sharply for examination. The specimens will gradually settle to the lower edge of the concavity until they touch both coverslip and slide. They can then be turned and positioned by moving the coverslip. Seal the slide with soft wax or vaseline to prevent evaporation.

Permanent Slides.   Fairly satisfactory permanent slides may be made with glycerin jelly in what is a standard procedure for many small invertebrates. Transfer specimens to a 10-precent glycerin and 70-percent alcohol solution in a concave slide and store in a dust-free place for about five days. The water will gradually evaporate and the glycerin permeate the specimens completely, thus leaving them in pure glycerin. Next, add a piece of glycerin jelly about the size of a wooden match head to a slide, and warm the slide over a flame to melt the jelly. Transfer a rotifer by means of a bristle into the melted glycerin jelly, lower a circular coverslip, and position the specimen, while observing with the compound microscope, by moving the coverslip from side to side. Never use a pipette when transferring specimens from glycerin (Harring and Myers, 1922), but rather use a mounted pig's bristle (a flattened toothbrush bristle glued to a matchstick will do; Pennak, 1953). A specimen may be balanced on the bristle and lifted from the glycerin for transfer. When the slide is thoroughly hardened, coat the edges of the coverslip with Murrayite cement or other waterproof and alcohol-proof cement. Label the slide with field and taxonomic data and store in a flat position.

Mounting Trophi.   Pennak (1953) suggests the following method:

Since genus and species identifications are often dependent upon the anatomical details of the trophi, it is sometimes essential that permanent or semi-permanent mounts of these jaws be made. The following method, requiring some skill and practice and much patience, is modified from Myers (1937). Place a drop of 1:10 Chlorox, or similar caustic alkali, just inside the concavity of a shallow concavity slide, and place a similar drop just outside the concavity. Then place a 22 mm. square coverslip on the outside drop and push it over the concavity until it is almost in contact with the inside drop. Next, place the rotifer in the inner drop by means of a bristle. If [the] coverslip is pushed slowly over the inner drop, the rotifer will be drawn under. And by working the coverslip over the concavity by short pushes, and adding small quantities of solution after each advance, the rotifer will be forced into the acute angle formed by the edge of the concavity and the undersurface of the coverslip. The edges of the coverslip should then be carefully dried and painted with Vaseline or Murrayite. The position of the rotifer should be noted and in about a half hour it will have dissolved except the trophi. Such trophi slides will last for several to many months, and if a Vaseline seal is used the trophi may be orientated by tapping or moving the coverslip slightly.*

Class Nematoda

The roundworms, class Nematoda, like the insects, protozoans, and vertebrate animals, have successfully invaded most available habitats. They are common in marine and fresh water, in soil or snow, as parasites or free-living animals. Barnes (1968) sums it up:

. . . they occur from the polar regions to the tropics in all types of environments, including deserts, hot springs, high mountain elevations, and great ocean depths. They are often present in enormous numbers. One square meter of bottom mud off the Dutch coast has been reported to contain as many as 4,420,000 nematodes. An acre of good farm soil has been estimated to contain from several hundred million to billions of terrestrial nematodes. A single decomposing apple on the ground of an orchard has yielded 90,000 round worms belonging to a number of different species.

Similarly, nematodes parasitize all manner of plant and animal material. Plant roots, bulbs, and stems are often so heavily infected as to render agriculture impractical. This writer has found nematode larvae and adults as common guests in the bodies of vertebrates and of large numbers of invertebrates, especially crustaceans. Chandler (1961, p. 365) sums it up by saying:

. . . probably every species of vertebrate animal on the earth affords harborage for nematode parasites, and Stoll (1947) estimated 2000 million human nematode infections in a world harboring 2.2 billion human inhabitants, a tribute, as he said, to the variety and biological efficiency of nematode life cycles.

Collecting Parasitic Nematodes.   Nematodes may inhabit any vertebrate host. Marine and fresh-water fishes, frogs, reptiles, birds, and mammals are all suitable hosts. However, unless a particular form of research is being conducted, limit classroom studies to laboratory animals or such vertebrates (fishes or frogs) as are prolific, abundant, and in little danger of being

* Robert W. Pennak, *Fresh-Water Invertebrates of the United States,* p. 186. Copyright 1953, The Ronald Press Company, New York.

overcollected. Look for parasites as soon as the host has been killed (see the appropriate chapter for methods of treating host animals). Open the body cavity of the host and, in turn, open all parts of the digestive tract. The stomach, intestine, and cloaca may all harbor adult worms. Occasionally the body cavity is invaded by adults and frequently harbors larval nematodes encysted as welts upon the internal organs (liver, spleen, digestive tract, body wall, mesenteries). Remove large worms first and then study individual sections of the host's digestive tract or other organs, submerged in an appropriate saline, for smaller ones (Appendix C). Worms should be transferred with a pipette or forceps to saline until ready for fixation. Also, remove the intestinal content to a dish, mix with saline, and examine under the dissecting microscope for parasites. Carnivorous and herbivorous invertebrates, as well as some filter feeders, should be examined in the same manner. Always keep nematodes isolated and labeled as to the part of the body in which they were found. Do not mix specimens from several parts of the digestive tract, as the exact locality of the collection is extremely important. Do not overlook the lungs and urinary bladder as likely sources of nematodes. Treat specimens as directed below.

Collecting Soil and Mud Nematodes.   Any damp soil around various types of vegetation; the mud bottom of ponds, lakes, and marine shores, sand from beaches, both above and within the intertidal zone or in freshwater areas; and soil beneath dung or other excreta harbor numerous species of free-living or larval, parasitic nematodes. The first method for separating specimens requires the least amount of equipment but the most work. Break up lumps of soil and dilute small quantities in a flat dish. Using substage lighting, hunt out nematodes and remove them with a pipette or fine dissecting needle (mount a size 000 insect pin in a match stick).

A second method of collecting nematodes is to wash the sample through a sieve made of window screen to remove coarse particular matter. Next, wash the material through brass testing sieves of 24 meshes, 50 meshes, 100 meshes, and if available, 150 meshes. (A fine plankton net may be substituted for the last diminsion.) Examine microscopically the washings from each successive sieve and remove specimens with a fine pipette.

The Baermann funnel is used as a standard tool for removing soil nematodes, is the most satisfactory, and requires the least amount of effort. To construct this funnel (Fig. 9–1) set up a standard ring stand with a supporting ring, place a large funnel in the ring, and equip the tip of the funnel with a piece of rubber tubing and a spring clamp or screw clamp. Next, fashion a piece of window screen to form a shelf that will sit part way down in the funnel, as shown in Fig. 9–1B. Fill the funnel with warm (50° C.)

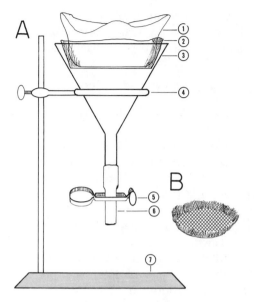

FIG. 9–1. The Baermann funnel for removing small nematodes. A. The funnel setup. (1) Cloth containing a soil sample, (2) screen shelf in place, (3) funnel, (4) ring stand, (5) clamp, (6) rubber tubing, (7) ring-stand base. B. The screen.

water so that it covers the screen by ½ inch to 1 inch. Place the soil in a piece of muslin or cotton cloth and set this on the screen so that it is covered by the water. Note, the screen may be eliminated by tying the soil within a cloth and suspending this in the funnel by means of a wire or string. After about ½ hour remove a small quantity of water into a dish by loosening the clamp. If the nematodes are numerous, withdraw only the bottom portion of water, but if they are few, centrifuge all of the water from the funnel to concentrate the nematodes. Treat specimens as directed below.

Collecting Plant Nematodes.   Plant nematodes may be removed with the Baermann funnel. These nematodes will be numerous in bulb or root nodules, which may appear partially decayed, from decaying vegetation, and the like. Bits of cut apple or other material may even be placed directly on the soil as a trap for such specimens. Macerate the plant material by teasing it apart with dissecting needles and then treat in the same manner (described above) as a soil sample.

Fixing Small Nematodes.   Very small to small nematodes (those which will conveniently fit under a coverslip) may be fixed in one of two ways. In the first method, pipette the specimens onto a sheet of glass with a minimal amount of water and fix by squirting them with hot FAA (60° C., Appendix C) which has ten parts glacial acetic acid instead of the standard two parts. This will increase the rate of permeation and, hence, fix the internal tissues more quickly.

The second general method is to pipette specimens into a concave slide depression and warm this above a burner flame until the worms stop wiggling. Do not boil or overheat. Remove excess water and fix by flooding with FAA.

Glycerin Jelly Mounts.   Mount as described for rotifers (p. 96).

Fixing Medium and Large Nematodes.   Specimens too large for slide preparation should be preserved in 5- to 10-percent formalin (depending on the size) or 75-percent alcohol. Smaller specimens may be dropped directly into FAA and transferred to the desired preservative, or they may be held between two glass slides and flooded with FAA. An alternate method for medium specimens, and a necessary one for large specimens such as *Ascaris,* is as follows: Straighten the specimens and place them on a piece of cheesecloth. In the meantime, bring a pan of water to a temperature near boiling. Roll the cheesecloth and grasp both ends with forceps; lower the worms into the near-boiling water, keeping them fully extended; and, after a moment, remove them. Next, place the specimens in a pan of FAA or 5-percent formalin to fix for 24 hours. Store in 5-percent formalin or run up through the alcohol and store in 75-percent alcohol.

Liquid Preservation.   Place specimens in appropriate-sized vials or jars with collecting data, including the name of the host, the collecting locality, and the place in the body from which the worm was taken. Vials should be cork-stoppered or fitted with screw caps and sealed in an airtight museum jar with an additional volume of preservative. Use 5-percent formalin (stronger, if worms are crowded) or run up through the alcohols and store in 70-percent alcohol.

PHYLUM ACANTHOCEPHALA

Collecting Procedures.   Acanthocephalans are routinely obtained during collecting of other parasites from the digestive tracts of vertebrates. In Washington, the common source is the Pacific tomcod (*Microgadus proximus*), which is heavily infected; also, occasional rock bass or ducks obtained by local hunters harbor acanthocephalans. Remove the visceral mass and carefully examine the stomach, small intestine, and large intestine by sectioning the digestive tract, placing sections in finger bowls with an appropriate saline (Appendix C), splitting the side of the tract carefully, and examining. Adults will have the proboscis firmly embedded in the gut wall. Usually the worms can be extracted by grasping them between the fingers and pulling gently. Other worms may dislodge themselves when left alone in the saline dish. Remove worms to dishes of clean saline where they will alternately contract and extend the proboscis. Leave them in the solution long enough to free the body of debris. A small water-color paintbrush can be used for cleaning. Be sure to keep specimens from various parts of the host's digestive tract isolated, as its location is significant in the ecology of the worm.

Fixing and Preserving. Always attempt to kill spiny-headed worms with the proboscis extended. After they are killed, specimens of all sizes may be preserved in liquid such as FAA or 5-percent formalin. However, smaller specimens are usually prepared on slides. Thus, in addition to ensuring that the proboscis is extended, fix these worms in a position such that they will fit under a standard coverslip.

Several methods may be employed to kill the worms with the proboscis everted. For small to medium-sized worms suitable for slide-making, transfer each specimen with a moderate quantity of saline to a sheet of glass. Position the worm so that it will fit under a coverslip. When the proboscis is everted, quickly drop a microscope slide on top of the worm. Usually, the suction created between the glass plate and the slide by the saline will maintain enough pressure to keep the proboscis everted. Otherwise, slight additional pressure could be applied to the slide during the killing. Prepare FAA as described in Appendix C, except that 10 parts of glacial acetic acid, rather than 2 parts, should be added. Put this solution on one side of the slide and pull it around the specimen by absorbing water on the opposite side with a piece of paper toweling. Maintain pressure on the worm during this procedure.

A second method is to place worms in the refrigerator until they are thoroughly chilled and expanded. (Owing to the cold marine temperatures in Washington, the author has found it necessary to further chill the worms by placing them in the freezer to render them insensible, being careful not to freeze the solution.) Next, siphon off the water and pour hot (50° to 60° C.) FAA solution over the worms.

A technique that works fairly well for very large acanthocephalans is to chill them as directed above. Then inject, with a fine hypodermic needle placed in the posterior part of the body cavity, sufficient FAA solution to extend the worm and force the proboscis to be everted.

Mounts with Permount or balsam and stained with Ehrlich's acid hematoxylin (following Cable, 1958) work very well. Cable recommends pricking the body wall with a fine pin to speed the movement of stains and alcohols into and out of the body. For this procedure wash specimens in 50-percent and then 30-percent alcohol for 30 minutes or more, depending on their size, and stain in Ehrlich's acid hematoxylin for several hours. Next, wash in 30-percent alcohol and move up through 50- and 70-percent alcohol, allowing at least 30 minutes for each washing. Finally, destain in 70-percent acid alcohol until the specimens are pink, and then wash in 70-percent alkaline alcohol until they appear blue. Finally, move specimens through 85-percent, 95-percent, and two changes of absolute alcohol for 30 minutes each, and clear and mount with Permount or balsam as directed in Appendix B.

REFERENCES

Barnes, Robert D., 1968, *Invertebrate Zoology*, Saunders, Philadelphia.

Cable, Raymond M., 1958, *An Illustrated Laboratory Manual of Parasitology*, Burgess, Minneapolis.

Chandler, Asa, 1961, *Introduction to Parasitology*, Wiley, New York. (Also published by Chapman & Hall, London, in 1957).

De Beauchamp, P. M., 1912, Instructions for Collecting and Fixing Rotifers in Bulk, *Proc. U.S. Natl. Mus.* 42:181–185.

Galigher, A. E., 1934, "The Essentials of Practical Microtechnique," published privately.

Harring, H. K., and F. J. Myers, 1922, The Rotifers of Wisconsin, *Trans. Wisc. Acad. Sci., Arts and Letters* 20:553–662.

Lavoie, M. E., 1958, The Preparation of Polyclad Whole Mounts, *Turtox News* 36(1):45–46.

Myers, F. J., 1937, A Method of Mounting Rotifer Jaws for Study, *Trans. Amer. Microscop. Soc.* 56:256–257.

Myers, F. J., 1942, The Rotarian Fauna of the Pocono Plateau and Environs, *Proc. Acad. Nat. Sci. Phila.* 94:251–285.

Pennak, Robert W., 1953, *Fresh-Water Invertebrates of the United States*, Ronald, New York.

Stoll, N. R., 1947, This Wormy World, *J. Parasitol.* 3(1):1–18.

Turtox Service Leaflet No. 2, 1959, General Biological Supply House, Chicago.

Wardle, R. A., and J. A. McLeod, 1968, *The Zoology of Tapeworms*, Univ. of Minnesota Press, Minneapolis.

The Higher
Worm Phyla

PHYLUM ANNELIDA

Class Polychaeta: Sandworms, Tube Worms, and Others

Collecting. Polychaete worms are found in and upon mud and sand, under and upon rocks, on pilings and floating objects, attached to marine algae, and as commensals associated with other animals. Thus, these worms occupy all types of beach as well as deeper water habitats. When collecting intertidally you will notice a marked difference in distribution of different species between the high- and low-tide levels. Therefore, all levels must be sampled if a complete faunal list is being made.

A sand screen is essential for collecting in mud and sand substrates. An ideal screen for this purpose may be constructed as shown in Fig. 10–1.

To screen for polychaete worms, place a shovelful of mud on the screen, wade into the water, and slosh the screen up and down. This will permit most of the sand and fine mud to go through the screen, but will maintain amazingly small worms long enough for them to be collected with a forceps. Standard sieves may be used in the same manner, using different sizes of screen for different sizes of worms. Select and screen mud from all levels and of all compositions of sand and gravel.

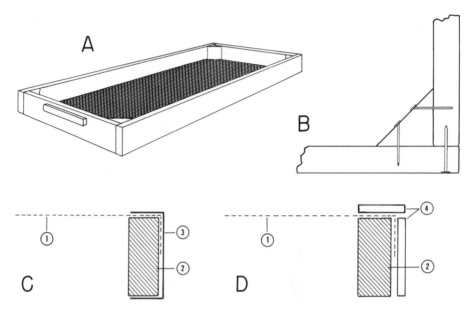

FIG. 10–1. Construction of a beach screen. A. Finished screen. B. Diagram for bracing and nailing the corners. C. The use of sheet metal to protect the screen. D. The use of wooden lath to protect the screen. (1) Screen, (2) side board, (3) sheet-metal guard, (4) lath.

Examine the undersides of rocks, found in all substrates, for free-living polychaete worms, and hunt through mussel beds, barnacle beds, and the like. Search through kelp holdfasts, as these harbor many species. Examine kelp fronds, coralline algae, and the like, for small tube-building polychaetes. The sides of rocks may contain numerous chitonous or calcified tubes of polychaetes. These must be removed with a pocketknife or chisel and hammer, or else the entire rock must be collected. The fouling organisms on pilings and floats are also likely hiding places for polychaete worms.

Narcotizing. Although researchers frequently preserve in bulk, specimens intended for display, classroom study, or even taxonomic work, should be thoroughly relaxed before preservation. The author has experimented with large numbers of the scale worm *Halosydna*, various species of *Nereis*, the shell-binder worm *Thelepus*, the iridescent worm *Hemipodia*, the plume worm *Eudistylia*, and the proboscis worm *Glycera*. Various narcotizing agents—clove oil, menthol, alcohol, epsom salts, Chloretone, Benzocain, chloroform, and others—were tried. Results are not always consistent, even with the better techniques, but some are much more reliable than others. Bear in mind that the volume of water containing the specimen will have an effect on the time and efficiency of the narcotizing agent. Specimens treated overnight are perhaps simplest to deal with, in that you do not disturb them by constantly probing and pinching. Occasionally, specimens are completely insensitive to severe pinching and probing but, nevertheless, contract when placed in formalin. Perhaps the best results are obtained by the alcohol method, fair results being obtained with the menthol, Chloretone, or clove oil method. When combined with the alcohol technique, menthol gives good results.

For the alcohol method, place specimens in a container of sea water large enough to permit expansion. Add 75-percent alcohol, drop by drop, until the worms become insensible. Stop the procedure when enough alcohol has been added to equal about 10 percent of the water or when the sea water becomes slightly milky. Test the worm by gently probing and, finally, by pinching the posterior end with forceps. If the worm continues to be active after the alcohol is introduced, test the worm each hour until the specimen is insensitive. When the worm no longer responds, stretch it out on a rough-sawed board and add a small quantity of FAA (Appendix C, use Bouin's fixative for histological specimens) to the specimen. After a few moments stretch the specimen out in a pan of FAA for 24 hours. Preserve as directed below.

Epsom salts give mixed results, some specimens being completely narcotized, whereas others periodically revive. One large plume worm in the author's collection revived off and on over a period of 72 hours, and finally died in a well-expanded condition.

The author finds that proboscis-bearing worms such as *Glycera* or *Nereis* often die with the proboscis fully expanded and the jaws exposed, when permitted to stand in a bucket of seaweed and other specimens in the laboratory. The oxygen is used up as the water putrefies. The addition of some organic material (the gut mass of a sea cucumber or any similar material) will hasten this process. Preserve specimens as soon as they are narcotized or dead.

Preservation and Storage.   If time does not permit anesthetizing, specimens should be placed in a bottle of sea water to which enough formaldehyde is added to make a 10-percent solution. After a week or more in this solution, they may be transferred to either 5-percent formalin, or, preferably, 70-percent alcohol. Some specialists strongly prefer 50- to 70-percent isopropyl alcohol for polychaete worms.

Small specimens are isolated in vials, along with their field data. The vials are filled with preservative, are cotton-stoppered, and are maintained in a large airtight jar also filled with preservative. Larger specimens are kept in a cool, dark place when not in use.

Class Oligochaeta: Earthworms and Related Species

Collecting Aquatic Specimens.   Ponds, lakes, and slow-moving streams and rivers (both above and below the water line) are favorite habitats for small oligochaetes. These worms live in mud and sand, around decaying vegetation, under the loosened bark of submerged logs and branches, under submerged boards or pieces of paper, and in similar habitats. Any of these objects should be carefully retrieved from the water by hand or with a dredge net (Chapter 12) and examined. The upper ⅛ to ¼ inch of mud or sand should be carefully scraped into a container and examined, a small portion at a time, under a dissecting microscope. Specimens may be maintained in aquariums in the laboratory for moderate periods of time.

Collecting Earthworms.   Worms are usually obtained by digging the surface layers of moist soil. The species composition is likely to change from place to place as the nature of the soil (grain size, chemical composition, pH, organic content) changes. Collecting in domestic areas will not necessarily produce a true picture of the species composition, in that many worms are transported into an area with plants. Digging near buildings, where grass is lush and green, under manure piles, and in other such areas will produce many specimens. Specimens may be taken at night following a very heavy, prolonged rainstorm or an all-day lawn sprinkling. Under such conditions worms come partially out of their burrows, but they are very sensitive to bright light and heavy vibrations. Nevertheless, they may be approached, grasped firmly, and extracted in great numbers. During the wet season any kind of ground litter, from old boards to rotten paper, may harbor earthworms. Electrical stimulators may also be used (see Fig. 10–2).

Narcotizing.   Various degrees of narcotization may be used for small aquatic oligochaetes. Specimens may be placed on a wet glass plate, permitted to crawl and thus stretch out, then killed with hot (60° C.) FAA or Bouin's fixative. By another method, they may be placed in appropriate containers of fresh water, supplied with a few drops of chloroform (which will sink to the bottom if introduced beneath the surface of the water), and covered with a glass plate until anesthetized. Alternately, epsom salts may be introduced in small, but increasing, quantities until the specimens are insensitive. Chloretone added initially to the container will eventually anesthetize the specimen. When the specimens no longer respond to mechanical stimuli, such as probing, siphon off the narcotizing solution and replace with FAA or 5-percent formalin.

Earthworms are best prepared by placing them in water and adding, drop by drop, sufficient alcohol to make a 10-percent solution. When fully anesthetized they may be fixed in 5-percent formalin. To prepare internal

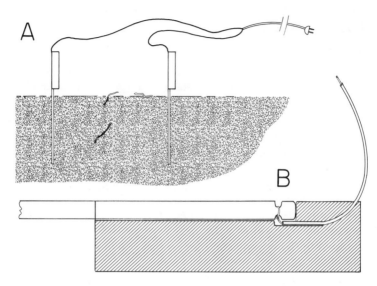

FIG. 10–2. Electrical shocker for collecting earthworms. A. Foot-long metal rods with tape-wrapped handles are placed in the soil, 3 feet apart, before the cord is plugged into a household electrical outlet. B. Cross-section of wooden handle before wrapping with electrical tape. The electrical shocker usually forces earthworms to come out of the soil voluntarily.

organs in the best manner for dissection, the Carolina Biological Supply Company injects earthworms with chromic acid, which hardens and thoroughly fixes the internal organs. The following instructions were kindly supplied by the company:

Inject specimens with a 1-percent chromic acid solution by introducing the hypodermic needle in a posterior direction about one inch behind the clitellum (the conspicuous glandular swelling). One injection will usually suffice to render specimens turgid. Next, straighten the specimens, roll them into a bundle, and submerge them in a 1-percent chromic acid solution for a period of four hours. Following this, wash the specimens in running tap water from twelve to sixteen hours, or until the chromic acid has been displaced, then run them up through various alcohol solutions of increasing strength and store them at 85-percent alcohol. Specimens may lose some of their color and become partially flattened by this process. However, many of the internal structures which are difficult to locate, such as the nephridia, will stand out with clarity.

Preservation and Storage. Oligochaete worms may be kept in 5-percent formalin or moved up through the alcohols to 75 or 85 percent. Store as directed for polychaetes.

Class Hirudinea: Leeches

Collecting.   Fresh-water leeches may be collected in large numbers in lakes, ponds, and slow-moving water, attached to the undersides of rocks, submerged structures, or aquatic vegetation. They are occasionally taken on host species which they parasitize, such as fishes, turtles, and frogs. Marine leeches are less numerous and are usually collected incidentally. The author has found them most abundant in moderately deep water, attached to rocks or other debris. Also, some species of marine fishes seem to habitually possess leeches, but, again, these are taken incidentally. Terrestrial leeches, it is often said, will find you if they are present.

Narcotizing and Fixing.   Small leeches may be placed on a glass plate and, when expanded, covered with a microscope slide. Immediately, FAA should be added to the edge of the slide so that it will run around the specimen. Do not flood the plate and thus float the slide away from the specimen. After 20 minutes move specimens into FAA, 5-percent formalin, or 70-percent alcohol. Many narcotizing agents for leeches are listed in the literature. Pennak (1953) suggests carbonated water, Chloretone, chloroform vapors, chloral hydrate, and other compounds. The author has always used Chloretone, as described for oligochaetes, with great success. Following narcotization, specimens may be placed directly in preservatives or fixed under slides with FAA or Bouin's fixative.

Preservation, Storage, and Slide Mounts.   For these topics follow the instructions given under the class Polychaeta, above.

MINOR PHYLA

Sipunculoidea

Sipunculid worms are moderately large animals found in sand or mud from the intertidal zone down to extreme depths or, in protected rocky coast zones, wedged between rocks which make up the substrata. The specimens are obtained by dredging (Chapter 1) or are collected by hand by turning rocks or looking for their tentacles protruding from hiding places. When collecting in rocky places remember that tremendous damage may be done to the general habitat if rocks are not returned to their normal position.

Echiuroidea

Members of this phylum inhabit muddy and sandy bottoms of tropical and temperate marine waters. They are sausage-shaped and may vary in color from greenish-gray to red. They possess a proboscis, but lack tentacles around the mouth. These worms are collected by dredging or by using the orange-peel bucket on appropriate marine bottoms (see Chapter 1).

Narcotizing and Preserving. These phyla are contractile and will withdraw the proboscis or tentacles, as the case may be, when preserved. They may be preserved in 5- to 10-percent formalin solution, but should be relaxed and expanded first. The *British Museum* collector's guide (1954) recommends menthol crystals or alcohol for narcotizing. Place the specimens in finger bowls containing sea water and sprinkle liberally with menthol crystals after they are fully expanded. After an hour or two begin to introduce, drop by drop, enough alcohol to make a 10-percent solution. Otherwise, begin directly with the alcohol treatment without the menthol. Very carefully touch the contractile parts after a few hours, to determine if the animals are still sensitive. Should they contract immediately, wash them in fresh, cool sea water, place them in a finger bowl of clean sea water until they are expanded, and begin anesthetizing all over again. Alternately, try Chloretone or epsom salts, as directed for polychaete worms above. Specimens may be stored in 80-percent alcohol or 5-percent formalin. Change the preservative solutions after 24 hours.

REFERENCES

Barnes, Robert D., 1968, *Invertebrate Zoology*, Saunders, Philadelphia.

British Museum of Natural History, 1954, *Instructions for Collectors, No. 9A: Invertebrate Animals Other Than Insects*, London.

Buchsbaum, Ralph, 1948, *Animals Without Backbones*, Univ. of Chicago Press, Chicago.

Carolina Biological Supply Co., 1963–1964, *Cataolgue No. 34*, Burlington, N.C.

Crowder, William, 1931, *Between the Tides*, Dodd, Mead, New York.

Knudsen, J. W., 1966, *Biological Techniques: Collecting, Preserving, and Illustrating Plants and Animals*, Harper & Row, New York.

MacGinitie, G., and N. MacGinitie, 1968, *Natural History of Marine Animals*, McGraw-Hill, New York.

Needham, James G., *et al.*, 1937, *Culture Methods for Invertebrate Animals*, Dover, New York.

Pennak, Robert W., 1953, *Fresh-Water Invertebrates of the United States,* Ronald, New York.

Pratt, H. S., 1935, *A Manual of the Common Invertebrate Animals,* McGraw-Hill, New York.

Storer, Tracy I., and Robert L. Usinger, 1965, *General Zoology,* McGraw-Hill, New York.

(See Chapter 1 for other references.)

CHAPTER 11

The Mollusks
and Brachiopods

A word of caution. Collecting mollusk shells has long been a popular hobby for private individuals and shell clubs. This in itself is excellent, and we must credit much of the classification of the mollusks and the compilation of faunal lists to such individuals or groups. However, the many books dealing with mollusks, catalogues and lists of shells for sale or exchange, and "nature guides" which tell some of the details of collecting and curating mollusks say little about the danger of overcollecting. Myra Keen (1971) is to be commended for her fresh statement urging sensible collecting and collecting procedures. Those of us who have been fortunate to collect in distant lands and on remote islands can easily detect the difference between the richness of fauna and flora there and on our own beaches. Thus, the reader is urged to collect only small series of specimens and to leave the habitat as unmolested as possible.

PHYLUM MOLLUSCA

Class Gastropoda: Snails, Slugs, and Related Forms

Habitats and Collecting

1. Marine collecting.   Gastropods are found in all marine situations. A tremendous wealth of specimens occurs intertidally. Low tides, therefore, are best for hand collecting. Night collecting, when safe (see Chapter 1), is often more productive than daytime collecting, especially in bays and estuaries, for most mollusks are nocturnal in their behavior.

When collecting on rocky beaches along the open ocean check some of the following places: (1) The spray zone above the high-tide level where species of limpets, turban snails, and others feed on the microscopic growths of algae. Look especially in small crevices where splashed water will accumulate and return to the sea. Moving down in the intertidal zone check (2) the tops, sides, and undersides of rocks for limpets and a host of other snails, turning rocks back to their normal position when finished. (3) In tide pools look among the fronds of algae and collect not only the large colorful snails but the minute species as well. Nudibranchs will be abundant in both the protected and partially protected rocky coast zones clinging to algae or other objects. Gloves and a geology pick are handy for handling rocks, and a dull knife is often essential to slip beneath the shell and foot of an unsuspecting snail, limpet, or chiton to pry it loose. See also collecting techniques mentioned in Chapter 1.

On sandy beaches in bays and estuaries a small amount of rock rubble will usually be present. One should screen sand and mud, collect around seaweed and rock deposits, and look for tracks of snails in the sand while wading out in the water. During a low tide, the moon snail and olive snail often remain buried but make horseshoe-shaped impressions or little furrows in the surface of the sand as they move underneath. A large number of oyster drills and whelks are also found in this habitat. Traps and baits, described below, may also be set for snails.

In the offshore waters skin diving and dredging are excellent means for bringing up deep-water mollusks. For dredging techniques and dredge construction, consult Chapter 1.

2. Fresh-water collecting.   Snails are found in all types of fresh-water situations. In lakes, ponds, and slow-moving streams they may be found on muddy bottoms, especially near the edges of the water, attached to rocks, sunken branches and logs or other debris, and dwelling on fresh-water plants. Other than direct observation, the best means for locating fresh-water snails is to collect various kinds of debris and examine them on the shore.

The dredge net and dip net (Chapter 12) are very useful for collecting plants and other debris. Carefully pull logs or branches from the water and examine both the upper and under surfaces, as well as under loose bark. Be on the lookout for the switchback trail left by the radula of a feeding snail or for gelatinous masses of eggs which signify that snails must be close by. Again, nocturnal collecting with a gasoline lantern and dip net is desirable, in that most snails are nocturnal. Specimens should be placed in jars of water and transported to the laboratory for proper preservation or culturing. Fresh-water snails may easily be kept in the laboratory aquarium when they are supplied with the proper food. Clams are found in many of the snail habitats described. Look for empty shells as a sign of their presence.

3. Terrestrial collecting.   The common land snail, *Helix*, is cosmopolitan in distribution; it may be found in very high mountains and deserts as well as in gardens and fields. Garden slugs and other species are also cosmopolitan. In the Pacific Northwest gigantic specimens of *Limax* make up part of the typical rain-forest fauna. *Limax* is almost always present during the daytime, except on days that are unusually dry or brilliant. On cloudy days these specimens can be found especially where leaf litter, fallen logs, and ferns and other shrubs are present. *Helix, Limax,* and related snails are also active at night and can be picked from vegetation, walls of buildings, sidewalks, along pathways in the woods, near streams or ponds, or in other damp places. Collect these species directly into a plastic bag with a small amount of vegetation to prevent crushing.

Preparation of Dry Shells.   The removal of animal tissue from the shells is undertaken (1) to permit easy storage and quick reference without the fuss of liquid preservatives, and (2) to eliminate the large number of jars and volume of alcohol that would be necessary to preserve all of the specimens in a taxonomic collection. Usually, smaller snails are preserved in liquid or placed in 5-percent formalin for 24 hours, then washed externally, and dried. Large snails are cleaned by boiling or by other methods.

1. Boiling technique.   Place specimens in water and bring slowly to a boil; then cool by adding tap water. The heat quickly kills the animal and loosens the flesh where it is attached to the shell. A wire basket made of ¼- or ½-inch mesh hardware cloth (Fig. 11–1A), or even a soup strainer, is indispensable for introducing and removing the specimens from the boiling water. Those snails with glassy, hard shells on the exposed surfaces may develop minute cracks; among these are the cowries, some of the *Murex* species with glassy shell openings, some of the cones and olive shells. Place these animals in warm water and very gradually bring them to a boil; cool slowly, never permitting them to contact cold water or cold objects until

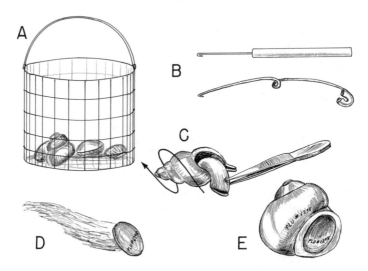

FIG. 11–1. Techniques of cleaning snail shells. A. Wire basket. B. Hooks. C. Twisting to remove soft parts. D. Operculum attached to cotton. E. Operculum glued in the cleaned snail shell with museum numbers intact.

they themselves are cool. Some of the very thin-shelled land snails should also be treated by this method.

When the specimens have been boiled and cooled, remove the soft parts by grasping them with a forceps or a hooked wire, crochet needle, or safety pin (Fig. 11–1B). With a firm grasp on the tissue, and on the shell, carefully pull the specimen out of the shell. Here, a little experience pays off in that you will develop a "touch." The tissue may adhere firmly to the shell, then suddenly become loosened, at which time you may jerk part of the animal free of the shell only to break off the soft terminal portion. Work close to some solid object, such as a table top, which your hand will strike the instant the tissue releases. Once the tissue has loosened, twist the shell in its normal direction (Fig. 11–1C). If part of the tissue remains in the shell, treat as directed below. Also, begin removing the radula as directed below.

Small snails may be treated by soaking specimens in either 5-percent formalin or 70-percent alcohol for 24 hours and then air-drying. Specimens are perfectly suitable for museum storage: the radula may be extracted later. Bergeron (1966) states that soaking dried gastropods in a trisodium phosphate solution (1 gram in 200 cubic centimeters water) not only prepares them for radula removal, but also causes the body proportions and color to return to a near life-like state.

A method extensively used at the Eniwetok Marine Biological Laboratory is that of placing fresh gastropods directly in sealed bottles containing a Pine Oil (brand) commercial cleaning solution, where they remain for a week or more, depending on size. The tissue is eroded by the cleaner and also decomposes. Yet, the pine scent masks any odor. Later, by shaking the shells under a jet of water, all remaining tissue is removed. This method is ideal for cowries, which have narrow shell openings, or other hard-shelled specimens, which may crack if boiled too vigorously. In Alaska, the author cleaned numerous gastropods by suspending them below a pier, in a fishnet bag, where amphipods and isopods removed all tissue in a matter of two days.

2. Retrieving broken flesh.   Should the "tail end" of the snail break off in the shell, this may be removed by filling the shell with water and shaking it vigorously, attempting to float the tissue out as the shell is drained. Directing a forceful stream of water into the shell sometimes will also dislodge this tissue. Many workers recommend placing formalin in the aperture and curing the tissue for 24 hours, rinsing the shell and drying it with the tissue intact. The procedure recommended for overcoming any resulting smell is to stuff cotton well back into the aperture of the shell. This technique, however, is not fully satisfactory, except for small specimens. The author prefers the method of placing the shell so that it will not tip over, and then introducing some 10-percent potassium hydroxide (KOH) down into the spiral. This will eat away the tissue. KOH will affect the brown outer layer (the periostracum) of some shells, but usually will have no effect on the other parts of the shell. After 24 hours, wash out this solution by directing water into the aperture, rinse thoroughly, and dry. This treatment may be repeated should you suspect that more tissue remains in the shell.

3. The operculum.   Many snails possess the operculum, which serves as a door and is attached to the side of the foot. When the animal retreats into its shell the operculum closes the aperture for protection. The operculum is either chitinous or hard and calcareous like the rest of the shell. The operculum must be saved, and should be removed from the soft parts of the animal at the time of cleaning. Keep the operculum with the shell while drying and cleaning. Record the field number or museum number on both shell and operculum; then glue the operculum to a piece of cotton (Fig. 11–1D), which is inserted in the mouth of the shell (Fig. 11–1E).

4. Removal of the periostracum.   Some collectors prefer to remove the outer, brown periostracum that occasionally occurs on snails and other mollusks. However, this layer is frequently ornamented, forming tufts of hair-like protrusions which in themselves are of taxonomic importance. While removal of the periostracum may be desirable to show the underparts of the shell, the practice should be restricted to only a few specimens in a series intended for taxonomic purposes. Remove the periostracum in 10-percent potassium hydroxide (KOH), or full-strength Clorox.

5. Coating shells. It seems to be a fad among amateur collectors to apply varnish or other substances to their shells. This creates an artificial appearance and should be avoided. The natural luster may be returned to the shell by rubbing it lightly with mineral oil (not olive or vegetable oil which may become rancid on the specimens) to enhance the appearance of the shell without altering the surface. By the same token, the treatment of shells in acid to produce a luster should be avoided, because acid alters the surface and lessens its taxonomic value.

Preparation of the Radula.   The radula is a rasp consisting of many rows of transversely orientated teeth. The number of teeth per row, their shape and occurrence can be reduced to a formula; thus, the radula is extremely useful for identification. The radular teeth are fastened to a belt-like sheet of tissue which draws back and forth over a cartilaginous "tongue" like a shoeshine rag pulled over the tip of a shoe. The radula is used to grind up plant and animal tissue, or even to drill through the shells of other mollusks.

The radula is generally mounted on a slide, in either a stained or an unstained condition. The radula may be freed from surrounding tissues by treating the head, or entire body if specimens are small, in a 10-percent solution of KOH or NaOH at room temperature for up to 24 hours; or treated in the same solution in a test tube (within a water bath) for about 15 minutes, allowing the water to almost boil. The radula is removed from its surrounding tissues under the dissecting microscope with fine forceps.

Perhaps the simplest technique for mounting is as follows: Wash the radula in several changes of water; transfer to a clean slide. Next, add a large drop of Turtox CMC-10 and as much CMC-S as clings to a dissecting needle. Remove any air bubbles, add coverslip with supports if radula is thick, dry, ring, and label (Appendix B).

Narcotizing Specimens.   Fully relaxed specimens are often required for study and dissection. All snails will contract badly when preserved under normal conditions. Some of the aquatic snails, however, are incapable of projecting very far beyond the shell aperture and relaxation is of little value. On the other hand, many snails, including land snails and slugs, nudibranchs, and sea slugs, can only be of value when fully relaxed prior to preservation. One must recognize that no single procedure is applicable to all snails, nor will a technique always work consistently, even with the same species treated under more or less the same general conditions. The procedures included herein, however, are usually quite satisfactory.

1. General techniques.   For many of the heavy-bodied, fresh-water and marine snails the following four methods yield some results.

Possibly the best general method is treatment with alcohol. Place the specimens in a container with their habitat water and add, drop by drop, enough alcohol to make a 10-percent solution. This should be introduced over a period of several hours—or days, if necessary. By very gently probing the expanded snail one can determine whether or not it will react in the preservative. When fully insensitive, the snail may be fixed in 10-percent formalin or FAA. Plunging specimens into concentrated glacial acetic acid gave only poor results for the author. Possibly the alcohol method can be improved upon in the following way: When the animal appears insensitive, slowly introduce small quantities of formalin so that it is gradually poisoned. As soon as it is dead, enough formalin should be added to make a 10-percent solution. Gradually introducing formalin without first using alcohol also gives fair results for some species.

An alternate method that is occasionally quite good, but somewhat difficult to gauge, is based on epsom salts. Place the specimens in containers of water from the habitat and slowly add crystals of epsom salts every 10 or 15 minutes, increasing the amounts after each hour. The crystals should be dropped in carefully, so as not to disturb the expanded specimens. Some specimens are easily relaxed in this way and may be transferred to 35-percent alcohol or 5- or 10-percent formalin for killing with no contraction. On the other hand, it has been the author's experience that many species require several days of treatment—and, even then, they have occasionally contracted.

Many workers suggest placing specimens in water and gradually heating them so as to kill them in an expanded condition. The author has found this method quite unsatisfactory in most cases.

The freezing technique of Gohar (1937), which he recommends for tritonid snails and nudibranchs, works quite well for many species of marine snails. (Fresh-water snails were not experimented with.) Place the specimens in a jar containing water, and when they are fully expanded, place them in the freezer. Most of the specimens will remain expanded and, thus, are killed by the reduced temperature. After freezing, chip away or melt away most of the ice and place the specimen, encased in ice, in strong formalin which will rapidly penetrate the body and fix the tissues. The author tried clove oil, Chloretone, menthol (recommended by many), chloroform, and other compounds, but with inconsistent results, both for fresh-water and marine (tritonid) snails. The large sea slugs should be treated either by methods mentioned above or as directed for nudibranchs below.

2. Narcotizing of Nudibranchia. The nudibranchs and related animals are possibly among the most beautiful of all the snails, and yet the most disappointing when preserved. These animals are brilliantly colored, adorned with external gills and other projections (cerata), and lack a shell. They display brilliant reds and oranges, blues, greens, yellows, opaque whites, or opalescent colors. Some of the Doris-type nudibranchs are fairly easily preserved, whereas the more highly ornamented forms such as *Hermisenda* (Abbott, 1954) or *Aeolis* are so delicate that they can scarcely be collected from the sea water without losing most of the cerata. Many workers recommend narcotizing with menthol and then (with or without narcotizing) plunging the specimens directly in concentrated glacial acetic acid. The author's experience here is quite disappointing, resulting only in badly contracted and dismembered specimens. For the more delicate forms the most promising techniques are the use of alcohol and then formalin, the gradual poisoning with formalin alone, or the freezing technique of Gohar, as described above under "General Techniques." The author has also obtained fair results by adding menthol to the water, covering the vessel for 12 hours, and then slowly adding formalin, a few drops at a time, over a period of many hours. Another method which gave excellent results in some cases, but was not fully consistent, was the use of Benzocain, a new local anesthetic available at any drugstore. A moderate pinch of this should be added to the culture jar. After 6 to 12 hours the specimen is usually completely relaxed and may be slowly poisoned by adding formalin, drop by drop, or plunged directly in 5-percent or 10-percent formalin. Five-percent formalin is an adequate preservative for nudibranchs; although alcohol will harden the tissues and make them somewhat more firm it will probably alter the color. See also MS-222, Appendix D.

3. Narcotizing of land snails and slugs. Many methods have been recommended for preserving of land snails such as *Helix* and land slugs such as *Limax*. One method found throughout the literature is that of placing the specimen on the end of a stick until it is fully expanded, then plunging it directly into boiling water. The author finds that even the best specimens prepared in this way are quite poor. The antennae are almost always contracted, the body shortens badly, and a tremendous quantity of mucus is discharged, indicating a significant reaction on the part of the animal. The author has experimented with menthol, clove oil, alcohol, chloroform, epsom salts, and direct injections with massive quantities of 10-percent formalin or FAA. Each of these methods gave only partially successful results. (After preserving more than a hundred large *Limax* simply to recheck the techniques from the literature, the author became so coated with mucus that a renaming of the land slug as "*SLimax*" seemed in order.)

The best technique is as follows: Place a moderate number of individuals in an appropriate jar, submerge the jar in a bucket of water, and place the cap on the jar (under water) so as to exclude all air. After 24 hours or more, the specimens can be preserved in formalin. Boiling the water beforehand is not essential, although slightly less time may be required for complete relaxation if the water has been boiled. Including some tobacco (such as a 2-inch length of cigarette used by Wells, 1932, and others) worked very well, but did not improve upon the plain-water method. The author placed six large *Limax*, measuring 4 to 7 inches, in a tall, straight-sided jar, and placed the jar on its side after the animals were sealed inside. Twelve *Helix* were similarly placed in a tall, straight-sided, half-pint jar. The *Limax* required 24 to 36 hours before they could be put directly in 10-percent formalin. It is best to test one or two specimens before the entire batch is pickled. *Helix* in previously boiled or unboiled water is almost always ready in about 24 hours. With this procedure, all of the animals expand fully, extend the antennae and the foot, and show no signs of local contraction.

Preserving Specimens. Specimens are frequently killed in the field in 5- to 10-percent formalin. Specimens should be washed in fresh water and transferred through 35- and 50-percent alcohol, and stored in 70-percent alcohol. Formalin solutions, unless neutralized, will act upon the shells and ultimately destroy them. Alcohol, on the other hand, will not harm the shell in any way, though it may act upon colors deposited in the soft tissues. No preservative will maintain animal colors adequately. Neutral formalin is perhaps the best preservative for maintenance of color.

Class Pelecypoda: Clams, Oysters, and Related Forms

Habitats and Collecting

1. Marine habitats. Pelecypods are found from brackish water to bays and estuaries, and on out to both the unprotected sandy beaches and unprotected and protected rocky beaches of the open ocean. Many of the species common to the protected and unprotected rocky beaches may also be found in large inland bodies of marine water, such as the Puget Sound Narrows, where currents are swift. Like the snails, clams and other pelecypods are distributed on the basis of the substrate and local conditions of the sea water. With every change in the nature of waves, currents, oxygen content in the water, temperature, and the like, there will be a change in pelecypod distribution. On exposed headlands and on pilings and rocks (Fig. 11–2A) mussels are found firmly attached by byssal fibers. Jingle shells are found lower down on rock surfaces, with one valve firmly cemented to the substrate. Frequently this inner valve may be loosened with

FIG. 11–2. Various pelecypod habitats. A. Mussels. B. Jingle-shell clam. C. Pecten attached to rock. D. Burrowing clam in rock. E. *Tagelus* in mud. F. Horse clam. G. Butter clam. H. *Teredo* burrowing in wood.

a knife, although occasionally a hammer and chisel will be essential for loosening the specimen. Pectins, other scallops, oysters, and the like, also attach to protected rocks. Look for holes in the rocks, especially on the sandstone formation, for many species of burrowing clams (Fig. 11–2D) may be collected with a geologist's pick or hammer and chisel.

Collecting is best at low tide. Specimens are usually dug by hand and located individually, either by the hole left in the sand by the siphon, or by an occasional squirt of water. However, sand, gravel, and mud may be screened with a coarse-mesh screen (built as directed in Chapter 10) to obtain small individuals.

2. Fresh-water collecting. Fresh-water specimens are collected by hand or by means of a rake or dredge net (Chapter 12). They can easily be located in the shallow waters of lakes, ponds and slow-moving rivers, and streams. Juveniles and small species are found associated with plant material. Such material is collected and carefully sorted through in white pans to locate the specimens.

Preparation of Shells.   Almost all sizes of pelecypods may be prepared as dried shells. Place specimens in fresh water and bring to a boil, continue to boil for 2 or 3 minutes (longer for large specimens), and cool slowly. Upon heating, the large adductor muscles usually become loosened, the shell opens automatically, and the soft parts are thus easily removed. The little muscle tissue that adheres to the shell may be pulled away with a forceps. At this point shells are generally closed and tied with string or fastened with tape while drying. One specimen should be left open, however, so that internal characteristics may be observed. Closed shells may be opened at any time with a short period of soaking in fresh water. The author prefers to leave fouling organisms attached to the shells; however, if the collector so desires, they may be removed with a pocketknife or by brushing with a stiff-bristled (not a wire) brush. Pennak (1953) recommends the application of varnish or Vaseline to the periostracum of fresh-water pelecypods. A little mineral oil may be used for other shells, but other coatings are not recommended (see p. 116).

Narcotizing and/or Preservation.   Marine pelecypods may be killed by soaking in fresh water. They may then be preserved in 75-percent alcohol after being moved through 35- and 50-percent alcohol. Another method for preserving all pelecypods is to place them in habitat water until the shells open. Next, insert a wooden plug $\frac{3}{16}$ inch thick, and preserve the animals by running them up through 35-, 50-, and 75-percent alcohol.

Class Amphineura: Chitons

Collecting Chitons.   Chitons are extremely numerous in the intertidal zone, but are restricted to solid substrates such as rocks or pilings. They occur from the splash zone into the subintertidal zone where they may be dredged. Chitons attach very firmly to their substrate, but may be loosened with a dull pocketknife or pulled free from the rocks by hand if they are taken by surprise. Specimens should be placed in a bucket of sea water with the foot flat against the bottom, in an attempt to prevent them from curling up.

Narcotizing and/or Preservation.   The main requisite for preserving chitons is to keep the foot and body flattened. This may be achieved by narcotizing or by binding and preserving. Most collectors simply place two specimens together, foot to foot, either in the field or in the laboratory, and bind these with strips of cloth or tape. String or rubber bands are not recommended as they distort the specimens. Specimens so bound may be dropped directly into 10-percent formalin or 35-percent alcohol for killing. All specimens are washed and treated in 35-, 50-percent alcohol, and finally are stored in 70-percent alcohol. When the bindings are removed specimens will remain flat and in perfect condition. Menthol, epsom salts, and the freeze-kill method have all yielded fairly good results with amphineurans, when used as directed for gastropods (p. 117). The author finds that specimens placed in clean sea water with several drops of clove oil added become relaxed in about ½ hour. If specimens move out of the water, simply return them until they are fully relaxed. Specimens narcotized by all methods had best be bound before preservation.

There are two methods of working with chiton shells. The shells may be dissected from the girdle, numbered with India ink (from anterior to posterior), and stored in small boxes. The second method is to preserve bound specimens in 10-percent formalin for 24 hours, wash in fresh water, and dissect away the main portion of the animal, leaving only the outer rim of the girdle to keep the shells intact. Such specimens should be kept in small boxes within insect-proof cabinets, inasmuch as dermestid beetles may attack the girdle.

Class Cephalopoda: Octopi and Squids

Collecting
1. Octopi.   The octopus usually has a home den, or hiding place, from which it ventures for feeding activities. These dens are below the zero tide level but may be exposed on extremely low tides. They are found where rocks are piled on top of one another so as to provide numerous spaces underneath. Any kind of container, such as a submerged tin can or bottle, should be drained and examined for small octopi. In Alaska the writer found large octopi (up to 30 pounds) at low tide by looking for trails of broken crab shells. Crustaceans make up a big part of the octopus' diet and, hence, are frequently scattered about the entrance of the hiding place. If the octopus can be reached by hand it is a simple matter to extract it, regardless of its size. Reach in and feel around until the ventral edge of the mantle is located. At this point a stout ligamentous muscle ties the mantle to the main part of the body. Insert a finger into the mantle on either side of this muscle and pull with a uniform pressure. The animal will soon release its grip and may be drawn from its hiding place. Larger octopi will not remain in a bucket, and should be placed in a plastic bag containing sea water

which is in turn carried within a bucket. Some collectors caution that octopi have a "poisonous" bite; thus avoid the mouth during capture.

2. Squids and other cephalopods. This collective group of animals is considered nectonic, that is, animals capable of directional swimming in spite of average currents. The squid, for example, is one of the few swimming invertebrates which follows a migratory pattern. Because of this, squids are somewhat seasonal in their occurrence at certain latitudes. In Puget Sound, for example, they are most frequently seen in the winter months from November through February, whereas in the Channel Islands off California they are very common during the summer months. Perhaps the best way to collect squid is with the use of a night light (see Chapter 1) set up from shipside in a remote anchorage. In Puget Sound these animals come in to the lighted fishing docks in tremendous numbers for a few weeks each year. Squids and like animals tend to remain on the surface, just beneath the night light, and can easily be dipped from the water. Some of these specimens swim toward the light with such velocity that they sail through the air at a height of 4 or 5 feet and land on shipboard. Another method that is always successful in the Puget Sound area is use of the beach seine, which is employed for catching fishes (see Chapter 16 for construction). A large 50- to 100-foot seine set 200 to 300 feet offshore on a sand and gravel bottom, and then skillfully retrieved to the beach, will usually yield many squids. Squids may also be obtained with large oceanic tow nets measuring 5 or 6 feet in diameter, but this equipment is rare insofar as the average collector is concerned.

Narcotizing and/or Preserving. Squids are best preserved alive. Place specimens in a large container (a bucket or plastic garbage can) big enough to permit swimming. Add enough formaldehyde to make a solution only a fraction of 1 percent. The amount is not important, so long as it is small. The specimens will swim in this, slowly die, and sink to the bottom in a natural position. Next, transfer specimens to the museum jars in which they will be stored, and position them as desired. Add 5-percent formalin for 24 hours, then replace with a fresh 5-percent solution. Place field data and taxonomic information in the storage container and seal with a lid. Small octopi and squid may first be hardened in plastic or wax-bottomed pans. Arrange the tentacles, body, and fins in the desired position, being mindful to bend the tentacles, if necessary, so that the specimen will fit the storage container. By drawing contracted tentacles through your fingers, you can stretch them out and they will remain elongated. Flood with 5-percent formalin for 24 hours, transfer specimens and their field data to appropriate bottles, add a fresh 5-percent solution, and seal.

Most of the nonshell-bearing cephalopods may be stored in 5-percent formalin, except for large specimens which may go into 8- or 10-percent formalin. Otherwise, the specimens may be transferred through 50-percent alcohol and stored in 70-percent alcohol. Formalin preserves the color better than alcohol, but is less agreeable to work with. Shelled specimens should be stored in alcohol.

Octopi and squids are easily narcotized. Bianco (1899) and others recommend the use of chloral hydrate, used in the same way as Chloretone. The author has had excellent results from the alcohol, epsom salts, and Chloretone methods. The first two techniques were conducted as described for snails (p. 117). All specimens were isolated in appropriate jars with sea water, placed in a semi-dark place, and narcotized as directed. Chloretone crystal was added initially, and again after 1 hour. Check specimens by gently probing the tentacles. Let them remain in the narcotizing solution for ½ hour after they show no reaction. For moderate-sized specimens, 1½ to 3 hours, or more, are required with alcohol; 3 to 6 hours are required with epsom salts, and 2 to 4 hours are required with Chloretone. Before flooding the specimens with a preservative, pipette a small quantity of preservative onto one tentacle. If there is a reaction place the animal back in the narcotizing solution. Should some faint contraction occur during preservation the tentacles may be stretched through the fingers and restored to normal length.

PHYLUM BRACHIOPODA:
LAMP SHELLS

Characteristics and Collecting.   The brachiopods are often confused with clams in that they possess two shells which cover the soft parts of the body. They actually resemble bryozoans much more than they do the pelecypods. For one thing, the shells are dorsoventrally located, and there is a median-posterior peduncle, a stalk which attaches the specimen to its substrate. Within the shells a lophophore with tentacles is supported on a shelly loop. These latter structures, and other anatomical features, are not characteristic of the phylum Mollusca.

Both the shell and soft parts are used as taxonomic features for identification. In the shell such things as the over-all shape, the cross section at various points, whether the two shells are articulated or not, the position of muscle scars, the peduncular opening, and the shape and location of the shelly loop are all of importance. Important features among the soft parts are the lophophore and its tentacles, the presence or absence of the anus, the location of various internal organs, and so on. It is therefore necessary to preserve specimens in such a way as to have expanded internal portions which may be readily examined.

Most of the brachiopods attach to solid substrates and are members of the lower intertidal fauna in the north latitudes, but are found in deeper (and thus cooler) waters as one goes toward the equator. Specimens have been dredged from depths of over 18,000 feet. Some genera, such as *Glotidia*, have a well-developed and expanded stalk which supports the animal on sandy substrates below the intertidal level.

Preservation. The minimal treatment a brachiopod should receive is as follows: Place the specimen in sea water until the shells are agape, insert a wooden plug between the valves, preserve in 35-percent alcohol, transfer through 50-percent and store in 70-percent alcohol. Formalin solutions damage the shell: if they are used, specimens must be washed as quickly as possible, soaked out in fresh water for several hours, and transferred up through the alcohols for final preservation.

A better treatment is as follows: Permit specimens to open and expand in sea water and add small quantities of alcohol, drop by drop, over a period of several hours until a solution of 5 to 10 percent results. Permit the animals to remain in this solution until they are completely insensitive, which may take 6 to 12 hours. At this time, test a single individual by placing it in alcohol. If it does not contract, the remainder of the specimens should then be preserved as described above.

STORAGE METHODS AND EQUIPMENT

Methods for Specimens in Liquid. Small specimens should be put into vials along with a label (printed in India ink and thoroughly dried) which gives the field data and scientific names of the specimens. Fill the vials with preservative, and stopper with cotton. Place these vials in a larger jar, along with some preservative and an internal label, and seal with an airtight lid. Keep specimens in a dark, cool place and check once or twice a year to make sure the preservative level remains constant. Larger specimens will go directly into airtight jars with their respective field and specimen data.

Unit Trays. Shells are kept in small cardboard unit trays along with a printed label containing the shell's number (which is also recorded on the shell), its name, and the field data. Small, conveniently sized cardboard boxes may be purchased at any large paper company, or they may be made from poster paper (lightweight glazed cardboard) to any desired dimension and depth, as illustrated in Fig. 11–3, A and B. Cut the cardboard into convenient squares of the right dimensions and place each in turn over a

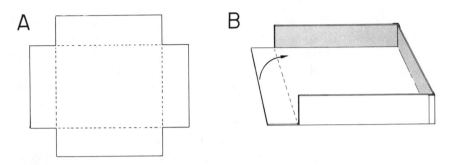

FIG. 11–3. Making individual shell-storage boxes of lightweight glazed cardboard and tape.

block of wood which has the corners cut out exactly as the finished cardboard should be (Fig. 11–3A). With a sharp pocketknife cut the corners out over the wooden template. Next, place the cardboard over a wooden block and bend the sides down. Finally, fold the sides and tape with masking tape (Fig. 11–3B). When specimens are kept in vials, these vials are in turn placed in such unit trays.

REFERENCES

Abbott, R. T., 1954, *American Sea Shells,* Van Nostrand Reinhold, New York.

Bergeron, E., 1966, *How to Clean Sea Shells: A Manual of Professional Methods for the Conchologist,* Marine Biological Research Associates, Balboa, Canal Zone.

Bianco, S. L., 1899, The Methods Employed at the Naples Zoological Station for the Preservation of Marine Animals, *Bull. U.S. Nat. Mus.* 39(M):1–37.

Gilbert, M., 1961, *Starting a Shell Collection,* Hammond, Maplewood, N.J.

Gohar, H. A. F., 1937, The Preservation of Contractile Marine Animals in an Expanded Condition, *J. Mar. Biol. Assoc.* 22(1):295–299.

Hyman, Libbie H., 1959, "Phylum Brachiopoda," in *The Invertebrates,* Vol. V, McGraw-Hill, New York.

Keen, A. M., 1971, *Sea Shells of Tropical West America,* Stanford Univ. Press, Stanford, Calif.

Knudsen, J. W., 1966, *Biological Techniques: Collecting, Preserving, and Illustrating Plants and Animals,* Harper & Row, New York.

Mattox, N. T., 1952 (Personal direction given in the Univ. So. Calif. Laboratory).

Morris, P. A., 1951, *A Field Guide to the Shells of Our Atlantic and Gulf Coasts,* Houghton Mifflin, Boston.

Morris, P. A., 1966, *A Field Guide to the Shells of the Pacific Coast and Hawaii,* Houghton Mifflin, Boston.

Muir-Wood, H. M., 1955, "A History of the Classification of the Phylum Brachiopoda," British Museum, London.

Pennak, Robert W., 1953, *Fresh-Water Invertebrates of the United States,* Ronald, New York.

Pilsbry, Henry A., 1939–1948, *Land Mollusca of North America,* 4 vols., Academy of Natural Sciences, Monograph 3, Philadelphia.

Storer, Tracy I., and Robert L. Usinger, 1965, *General Zoology,* McGraw-Hill, New York.

Thompson, J. Allan, 1927, "Brachiopod Morphology and Genera (Recent and Tertiary)," Wellington, New Zealand Board of Science and Art, Manual No. 7.

Verrill, A. H., 1950, *Shell Collector's Handbook,* Putnam, New York.

Webb, W. F., 1935, *A Handbook for Shell Collectors,* Lee Publications, Wellesley Hills, Mass.

Wells, M. M., 1932, *The Collection and Preservation of Animal Forms,* General Biological Supply House, Chicago, Ill.

(See Chapter 1 for other references.)

12

The Insects

Perhaps the sheer number of known species of insects is sufficient to explain the uncommon interest man has always had in this group of animals. Estimates of the number of species of insects range on the conservative side from 750,000 to over 1,000,000 species. The variety of insect morphology, color, and behavior alone is fascinating. Probably more serious biologists in all disciplines first gain their insight by the simple experience of collecting butterflies than by any other means.

Today, insects are seriously decreasing in numbers. Pesticides and other toxins probably destroy more insects per year than do insect predators (which in turn are poisoned). Noncollecting is not the answer to insect survival. However, collectors should treat each specimen as being extremely valuable; therefore, keep good field notes, label faithfully and fully, and mount and store specimens properly. Collections no longer wanted should be turned over to museums of natural history.

Building an insect collection includes the steps of field-collecting, mounting or preserving, identifying, and storing and caring for specimens. This chapter will detail the techniques for building and using the equipment needed and will provide information for all levels of insect study.

COLLECTING METHODS
AND EQUIPMENT

Basic Net-Making Techniques

A simple net consists of a wire ring or hoop, a cloth bag of some sort, and a suitable handle. Usually, instructions for the construction of a net begin with exact measurements for the hoop diameter, the length of the cloth bag, the size of the handle, and so on. Too frequently students give up attempts to build a net because of the rigid measurements imposed in matching the bag diameter to that of the hoop. Nets can be constructed very cheaply and very easily without resort to a complex formula of measurement. In brief, all one has to do is to sew the bag to any rough dimension that he chooses, then fit the wire hoop into the bag, and finally, attach the hoop onto some suitable handle.

Many kinds of bag fabrics are available locally or through biological supply houses (see the General Biological Supply House catalog for fabric illustrations). The fabrics most frequently used for entomological work are cotton, nylon or dacron marquisette (this is always obtainable in dry goods or catalog stores), nylon or cotton bobbinet, or Brussels netting of a cotton or nylon grade. The specifications and uses of these various materials will be discussed below under specific kinds of nets.

After the netting is selected, determine the approximate hoop diameter you want. This may be 8 or 10 inches for dip nets or ground nets, 12 to 15 inches for aerial nets, and so on. In cutting and sewing of the fabric, the circumference of the net bag should be approximately three times the diameter of the hoop; the depth should be from one to two times the hoop diameter, depending on the particular kind of net being made. For example, if the hoop diameter is to be approximately 10 inches the bag circumference is 30 inches and the depth 15 to 20 inches. After the fabric has been cut, fold a 1-inch hem along the top edge of the bag, as illustrated in Fig. 12–1A. This hem will receive the hoop and thus should be stitched at least twice on the sewing machine to ensure that it will hold. If the net is to be used around brush or in water it is a good idea to make the hem of a stout fabric such as muslin, white duck, or canvas. Figure 12–1I shows a simple hem made of canvas and attached to the netting fabric. Figure 12–1J shows a hem made of canvas with an additional apron attached along the outside of the net to protect the net from weeds or other obstructions. Once the top hem is complete, fold the cloth double and stitch the outline of the net bag as shown in Fig. 12–1B. Particular attention should be paid to stitching the bag around the bottom curvature, for frequently the meshes will tear loose

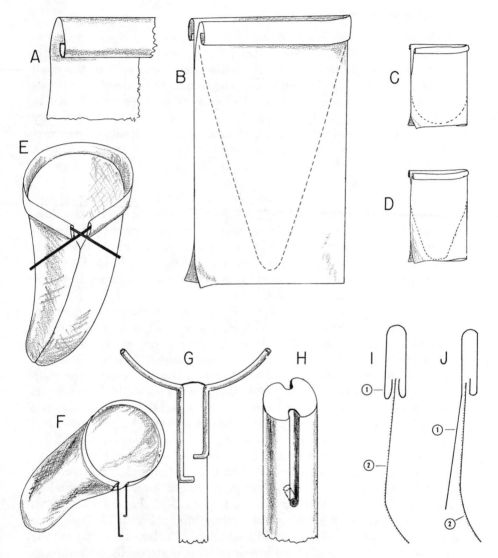

FIG. 12–1. Net-making technique. A. Folding the hem. B. Stitching the bag. C. Aerial net pattern. D. Sweep-net pattern. E. Forming the hoop. F. Finished hoop. G–H. Attaching hoop to handle. I. Reinforced hem. J. Reinforced hem wtih apron. (1) Canvas reinforcing, (2) net material.

at this point. Stitch the bag two or three times to ensure holding the fabric. Figure 12–1C shows the outline of the bag used for aerial nets; Fig. 12–1D shows the outline used for sweep nets. Trim away the surplus cloth around the bottom of the net bag, leaving at least ¼ inch of fabric to prevent raveling.

Once the net bag is completed, a stout wire is introduced into the hem at the top of the bag. Number 8- or 9-gauge galvanized wire, available in hardware stores, proves very suitable for most kinds of nets. Round off the end of the wire with a file to prevent snagging. Once the wire is part way into the hem it may be necessary to bend the wire slightly, but smoothly, so that it will round the curvature of the hem, as shown in Fig. 12–1E. Cut the wire off, leaving at least 6 inches projecting from each side of the hem. Next, clamp the wire into a vise and bend it, as shown in Fig. 12–1, F and G. Finally, bend the ends of the wires inward, at a 90° angle, to form "ears" which will project into the net handle about ⅜ inch.

Although any suitable stick (willow, etc.) may be used for a handle, a ⅝-inch hardwood dowel (¾ inch at the maximum) proves to be the most suitable for general aerial nets. The exact length and diameter of the handle will be determined by the intended use of the net, however. For example, a long, stout handle is required for a dip net. The handle must be as straight as possible in order to give the net the proper balance. If your net handle has a bend, place the bend at right angles to the plane of the hoop to prevent the net from twisting in your hand. Place the hoop alongside the handle and mark the location of the wire "ears." Next, drill holes for the ears and cut a partial or complete groove along the side of the handle from the hole to the tip (Fig. 12–1H). Finally, attach the hoop to the handle by placing the wires into the grooves and taping the hoop in place. Electricians' plastic tape (not rubber tape or friction tape) should be used to tape the hoop in place. This completes the net, except for adjustment of the handle to its proper length. If your net is an aerial net the handle should be fairly long, but not so long that balance and maneuverability are impaired. Therefore, test the net out of doors and, if it seems too heavy and clumsy, reduce the length of the handle somewhat. This will ensure a greater catch of insects in the field, in spite of the reduced length.

Specialized Nets

Aerial Net.   Most winged insects that are active during the warm summer months must be captured with an aerial net. Build this net as described above, using a hoop diameter of from 12 to 15 inches. Any of the marquisette fabrics are very useful, though the nylon fabric will stand up under wet conditions somewhat better than the cotton. The nylon, however, is more difficult to stitch and often very difficult to dye if special colors are desired. For collecting butterflies the lightweight nylon fabrics are desirable in that they do the least amount of damage to the wings. The depth of the bag should always be 1½ to 2 times the diameter to ensure that when the bag is closed it will still provide ample space for the specimen at the tip. The bottom of the bag should have a rounded curve rather than coming to a point, in order to prevent butterflies or other fragile insects from becoming lodged and damaged at the tip. The length of the net handle should depend on maneuverability rather than on the over-all reach attained. However, in the tropics and elsewhere extremely long handles are often necessary to reach butterflies and other insects that are high above the trail. Sections of aluminum tubing which can be joined together, may be made from the legs of an old camera tripod.

Sweep Net.   The sweep net is designed to be used in heavy vegetation and, therefore, should be constructed of muslin or white duck or light canvas. The bag hoop and handle are constructed as described above, the bag being more pointed, however, as illustrated in Fig. 12–1D. The sweep net is fashioned with a fairly short, stout handle. The collector sweeps it back and forth through the grass or bushes as he walks (Fig. 12–2K). This net collects a tremendous number and variety of small insects in a very short time, and therefore the contents of the net should be checked frequently and the specimens sorted and placed in proper killing vials. Sweep netting is usually very good on the third or fourth warm sunny day in spring and will continue to be quite good during most of the summer period.

Umbrella, or Beat, Net.   The umbrella, or beat, net has proved to be an invaluable collecting tool which is too little used by most collectors. It is especially useful for collecting the many species of insects that move up onto higher bushes and trees during the spring and summer months. To make the beat net (Fig. 12–2, A and I) some stout cloth such as muslin or very light canvas should be used. Cut a piece of cloth 42 by 42 inches and sew a double 2½-inch hem on all margins. When this is completed, a square 32 by 32 inches will be obtained. The corners may be reinforced with leather in order to receive the cross sticks (Fig. 12–2C) or provided with tie strings as shown in Fig. 12–2D. Next, cut two sticks or dowels long enough to go from corner to corner. These may be tied together in the center or bolted together with a ⅛-inch bolt and wing nut. One of these sticks may be wrapped with tape to provide a stouter grip and thus make the net more maneuverable. It is often desirable to reduce the length of the

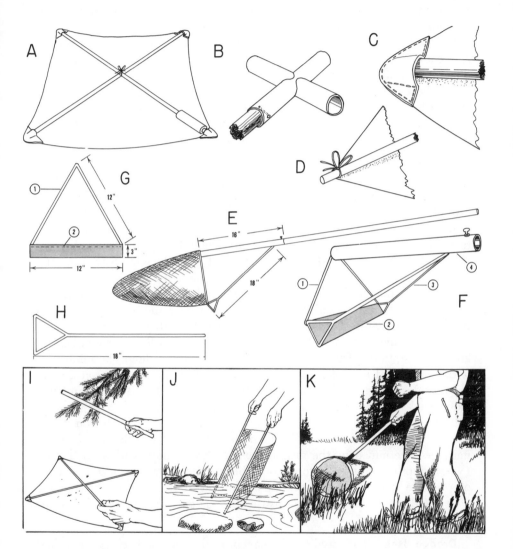

FIG. 12–2. Some additional nets. A. Finished beat net. B. Central cross. C–D. Corner construction. E. Completed dredge net. F. Frame. (1) Triangular hoop, (2) rake, (3) brace, (4) metal tube with thumbscrew. G. Triangular hoop. H. Side-brace. I. Using beat net. J. Using aquatic-dip net. K. Using sweep net.

sticks so that they may be packed into knapsacks. This may be done by making a cross of ½-inch metal tubing (not pipe) of the variety used in plumbing water faucets. Cut the tubing about 2½ inches long and weld or braze these together in the center. Cut the crossbars to half the original length and fit them into the cross. If one of the sticks is wrapped to produce a handle, this should be permanently riveted into the metal cross to prevent it from rotating. With this kind of construction the net will be fairly light and compact and easily transported into high mountain country or other difficult terrain.

The beat net is used by holding it beneath protruding branches of trees and bushes and beating upon the bushes with some stout stick, as shown in Fig. 12–2I. The handle of a sweep net or aerial net makes an excellent beating stick and will thus provide an alternate net should a butterfly come into view.

Aquatic-Dip Net.   The aquatic-dip net is extremely useful for collecting many kinds of plant and animal life. A net fabric should be selected to provide the best mesh for the job desired. For example, the dip net is very difficult to move rapidly through the water, and speed is often very important. Therefore, the largest mesh obtainable should be used, provided it will still hold the organisms which are being captured. One-eighth inch or one-quarter minnow netting is therefore much more desirable for catching amphibians and reptiles than would be marquisette or bobbinet which are best for insect and general collecting. The depth of the net should be approximately equal to the diameter of the hoop. Slightly deeper nets may be used if extremely active and agile specimens such as fishes or amphibians are being captured. The hem at the top of the net bag should be reinforced with canvas and it is often a good idea to provide a canvas apron around the outside of the net bag, as illustrated in Fig. 12–1J. Usually number 9 galvanized wire is stout enough to make an adequate hoop, but for vigorous collecting especially among weeds and rocks a hoop of ¼-inch iron dowel is better than the lighter wire. The length and diameter of the handle again must be determined by the distance at which one must work with the net (water depth and other factors should be considered here) and the amount of pressure that will be placed upon the net as it is swung through the water. Otherwise, the general instructions given on page 129 should be followed.

Dredge Net.   The dredge net has proved to be a very popular and valuable net for collectors working in streams, ponds, lakes, and even in marine habitats. Many designs are available on the market. The dredge net shown in Fig. 12–2E is a net designed by the author for most kinds of aquatic collecting. The special feature of any dredge net is that it has a lip which can be dragged along the bottom, in the manner of a larger biological dredge, to pick up both attached and motile plants and animals.

The net (Fig. 12–2) is constructed by making a triangular hoop of $\frac{1}{4}$-inch metal rod. This is welded to a 3-inch piece of bar iron to serve as the metal rake. Two side braces are constructed of $\frac{1}{4}$-inch metal rod, as shown in Fig. 12–2H, and then attached to the dredge, as shown in Fig. 12–2, E and F. Cut a 16-inch piece of metal tubing or light pipe, of a diameter suitable to receive the net handle, and weld this to the top of the triangular hoop and to the two side braces. Provide the tubing with a wing bolt at one end so that the handle may be removed when the net is not in use. The net bag may be fabricated from metal screen (hardware cloth), plastic window screen, or any suitable cloth netting such as bobbinet or Brussels netting. The size of the mesh in the net will have to be determined by the specimens being sought. The more flexible cloth or plastic netting is much more desirable than the rigid screen netting, even though it may be necessary to protect the cloth netting in some way to prevent snagging. This net or any other net used in the marine environment must be carefully washed with fresh water and hung up to dry to prevent rusting or rotting.

Stream Net.   Figure 12–2J shows a very simple but useful net that may be used in streams in place of the dredge net. This is constructed by attaching either plastic window screen or cloth netting to two handles. These handles may either be held as illustrated, while another collector dislodges organisms upstream and allows them to float into the net, or the handles may be sharpened so that they can be driven into the substratum, thus permitting one collector to operate the net.

Other Collecting Devices

Aspirators.   The aspirator is an invaluable piece of collecting apparatus when small, fragile insects or other arthropods are being sought. Figure 12–3, A and B, shows two popular designs. Plastic (preferably) or glass vials or tubing should be used for the main body of the aspirator. In Fig. 12–3A a two-hole rubber stopper is provided with two metal, plastic, or glass tubes. One tube is used to receive the small insect; the other is provided with a piece of surgical tubing about 20 to 24 inches long with a piece of gauze tied or taped onto the opposite end of the tubing. The second model (Fig. 12–3B) is essentially the same, except that two one-hole rubber stoppers are used in a piece of plastic or glass tubing. The collector holds one end of the rubber tubing between his teeth and sucks in quickly when the collecting tube is placed near some small insect. The insect is

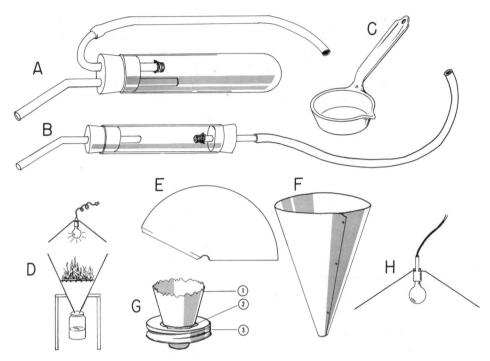

FIG. 12–3. Additional collecting equipment. A–B. Aspirators. C. Dipper. D. Berlese funnel in operation. E. Pattern for funnel. F. Completed funnel. G. Attachment of jar lid. (1) Base of funnel, (2) solder, (3) jar lid. H. Reflector assembly.

automatically pulled back into the collecting chamber and is held there until transferred to a killing bottle. The gauze prevents these insects from being inhaled by the collector.

Dippers. In some forms of aquatic collecting the dipper proves quite handy for obtaining small specimens such as mosquito larvae and pupae. Anything from a kitchen strainer to a regular dipper provided with a piece of window screen on one side will suffice. The dipper is convenient in that specimens may be carefully worked into the collecting chamber without unduly disturbing the bottom debris in which they hide.

Berlese Funnel. The Berlese funnel provides the best method for removing very small insects or other animals from litter and debris, such as leaf litter, coarse humus, nesting material such as bird nests or rat nests which might harbor mites and fleas, and so on. There are dozens of modifications for the Berlese funnel, but all of them consist of the same general parts. A fairly large metal funnel (illustrated in Fig. 12–3D) is equipped with a collecting bottle at the bottom, a shelf for holding the debris inside, and a source of heat above. The principle upon which this funnel functions is that the heat slowly drives all of the motile organisms down through the debris, either by overheating or by drying out the litter, and thus forces the specimens to fall through the screen and into the collecting bottle at the bottom. The writer has used this device to process hundreds of rat nests in order to obtain fleas, mites, and pseudoscorpions.

The diameter (14 to 20 inches) and the height (16 to 30 inches) of the funnel must be determined by the nature of the material to be processed and the approximate volume that must be handled at one time. Twenty-eight-gauge galvanized iron or other suitable metal should be cut into a half circle (Fig. 12–3E) and, after trimming, should be rolled and the sides either riveted or soldered together (Fig. 12–3F). Next, solder the outer metal ring from a fruit-jar lid to the lower side of the funnel (Fig. 12–3G). Construct some sort of stand for the funnel to prevent tipping. Finally, attach a light socket to a commercial light reflector or a shallow funnel (made in the same manner as described above). Suspend this above the funnel or rest it directly on top of the funnel. Experimentation may be necessary to select the proper size of light bulb. Factors to consider are the over-all dimensions of the funnel, the volume of litter being processed, and the amount of speed required in processing. Light bulbs that are too large are unsatisfactory in that specimens are killed before they can move down through the litter. It is better to take several extra hours, or even days, in processing than to attempt to rush the job. The collecting bottle may be dry or provided with some killing agent such as alcohol or FAA if the specimens are to be preserved.

The collector goes into the field and picks the kind of debris to be examined and brings it back to the laboratory or campsite for processing with the Berlese funnel. Plastic bags or paper bags are very suitable for picking up litter such as leaves, rat nests, and so on. Tightly seal the bag at the top by tying plastic bags with string or by rolling the tops of paper bags and stapling them shut. A slip of paper with all the necessary field data should be included in each bag; this should indicate the locality, date, the time of collecting, a description of the habitat in which the collection was made, the temperature and other environmental conditions, and any other data that may prove of value in research. Process the bags of litter as soon as possible to prevent specimens from dying inside. If there is some delay, be sure that the bags of litter are not subjected to overheating, dehydration, or freezing. Occasionally, as long as three days may be required in the total processing of material, and specimens will continue to fall into the collecting jar over that period of time.

Insect Trapping.   Many kinds of insects are conveniently collected in large numbers with some form of insect trap. Any material that is of some peculiar significance to the insect may be used as bait. For example, many kinds of food materials or food plants used by the larvae of certain insects will lure the adults and thus may be used as bait. Chemicals which produce odors resembling those of mating insects may also be used. Lights, and even mechanically produced sounds (e.g., to attract mosquitoes), have been used very successfully. In order to devise a suitable trap and a bait the habits and requirements of the particular species must be fairly well studied by the student. For example, fresh or rotten meat or vegetable matter will lure many kinds of beetles, flies, bees and wasps, and other forms. Cow dung or other forms of dung will lure many flies, beetles, and (in the tropics) butterflies, as well as other insects. Pine pitch and fir pitch, xylene, turpentine, or other aromatic compounds have been painted on fence posts, log stumps, and so on, to attract beetles, flies, wasps, and many other insects that are attracted to freshly injured trees. Sugar solutions, molasses, stale beer, and even perfume scents have been used to attract moths and butterflies, as well as other insects. Rotting fruit or alcohol will attract a whole host of species. Students interested in constructing some of the many kinds of specific traps not described herein should see *Entomological Techniques* (Alvah Peterson, 1959).

Tin-Can Traps.   When camping and collecting in a given locality entomologists frequently bury tin cans around their campsite in order to trap and collect beetles and other insects. Figure 12–4A shows a typical setting of a tin-can trap. A tin can is buried and baited with some suitable bait such as meat scrap, fat, or peanut butter. A board or piece of bark or even a flat stone should be placed over the top of the tin can to keep rodents or dogs out of the trap and yet provide enough room for the insect to enter. Traps of this nature may work for weeks at a time and can be visited daily by the collector.

Funnel Trap.   The author has devised a funnel trap for catching carrion beetles which are attracted to dead and decaying animals. Figure 12–4B shows a cross section through the trap. A large hole is drilled on either side of a wooden box and a screen funnel is tacked into this hole. One of the boards on the roof of the box is removable and serves as a door. The trap

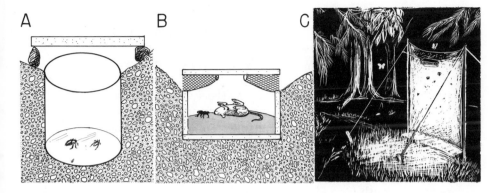

FIG. 12–4.   Insect traps. A. Tin-can trap. B. Funnel trap with bait. C. Night lighting for moths.

is partially buried, as illustrated, and is provided with 2 inches of dirt on the bottom. A suitable carrion bait such as a dead mouse is placed in the trap and a few drops of scent are also placed in the trap. The scent is prepared by rotting some meat or fish in an airtight jar which is half filled with water. After this has rotted for about two weeks it makes an excellent scent to be used if the bait provided is too fresh. This trap is much more elaborate than is needed to collect carrion beetles locally, for by placing a bait under a board or a stone one can obtain large numbers of carrion beetles. The trap is designed for long-range field trips where large numbers of traps are set out along the side of the road and retrieved three or four days later.

Light Traps.   Many kinds of light traps can be devised which will attract nocturnal insects. Some of these are provided with holding chambers and even with killing chambers so that they are essentially self-running. The moth sheet (Fig. 12–4C) is perhaps the most interesting of the light traps, but it must be attended by the collector. In principle a moth trap can be made by using any white surface (e.g., the side of a building) and a suitable light. However, in the field some sturdy fabric such as muslin, white duck, or even strong bed sheeting will do very nicely. The dimensions of the sheet should be approximately 6 by 8 feet. The sheet is suspended between two poles 6 feet high and 6 feet apart. This allows about 2 feet of the sheet to be stretched out as a ground flap.

A suitable light source is required to attract the insects. Pressure gas lanterns have often been used with success. These may be improved somewhat by staining the lens of the lantern to provide a blue or almost "black light" quality. By far the best light source is the 15-watt black-light neon tube. The light fixture and black-light tube can be obtained for only a few dollars and may be run from either alternating current or direct current. Generators, automobile batteries, or long extension cords will provide adequate power. The light should be suspended slightly in front of the upper margin of the sheet, thus illuminating the sheet and ground flap.

Collectors using light traps in the late spring and summer months will notice that the species composition differs from hour to hour during the entire night. Unless the temperature becomes unfavorable, one may collect the whole night through and get new and different species every few hours. The location of the moth sheet makes a lot of difference, for the species will vary tremendously according to whether you are collecting near water, in woods, out on an open meadow, or on a desert hillside. The collector will sometimes need five or six cyanide bottles so that no one bottle becomes too full of living specimens. When a reasonable number of moths or other insects have been placed in a given bottle, the bottle should be set aside until the specimens are dead. Specimens may be transferred to holding boxes, prior to field pinning or papering, thus freeing the killing bottles for new specimens. Do not mix beetles or other robust insects with moths, but instead kill them separately. On field trips the chlorocresol method of field storage is an extremely time-saving technique.

General Collecting Sites

Insects are available almost all year long, except when winter conditions are extremely severe. Large numbers of insects overwinter under stones, logs, leaf litter, or down in the ground, and may be found in the early spring in any likely situation which will have afforded winter protection. In the late spring and early summer many species metamorphose from juvenile stages and emerge as adults for the first time. In the summer, breaking old logs and looking under boards is less profitable than during the early spring or late fall, since only nocturnal specimens will be found. Toward the end of summer another new wave of adult insects metamorphoses from the larval stage and appears in time for fall hibernation. When the collector learns the general life histories of insects, along with their daily behavioral patterns, he can usually predict where the best collecting will take place.

Daily weather conditions have a lot to do with the availability of insects. Cold weather will inhibit activity and will prevent insects from leaving their hiding places or will keep them inactive once they have left their hiding places. Rain will usually drive insects down out of the vegetation and cause them to hide. The first warm, sunny day after a cool spell will produce a flurry of flying insects. However, several days may be required before those specimens in well-protected hiding places are warmed and activated.

The most productive collecting area is the ecotone. The ecotone is defined as that distinct area where two habitats overlap. For example, collecting is always better where the habitats of a stream and a meadow overlap, or where the meadow and the forest habitats overlap, since species from both habitats will be present. The seasonal appearance of many species is geared to the growth and development of plant populations, and thus any given habitat will have a year-long succession of plant and insect populations.

When collecting in the woods, turn stones and logs or remove loose bark from trees to find overwintering and nocturnal specimens. Tree trunks should be carefully examined and sweep or beat nets should be applied to bushes and low trees. Any sunny glade within the forest should be given particular attention, as this represents an ecotone. Mushrooms or other fungi will harbor both adult and juvenile insects. On warm days sap flowing from injured trees will attract a continuous succession of beetles and other insects.

In meadows the scarcity of hiding places makes turning any object on the ground worthwhile for the collector. The sweep net should be applied to low grass and bushes. If flowers are available, the aerial net will be of extreme value. Fungus, dead animals, or cow dung should always be investigated for the dozens of probable species that will be found there.

In streams and ponds collectors will find a continuous parade of insects, changing with the seasons. The dip net and dredge net will prove valuable in both day and night collecting. Bottom debris should be collected and examined on the shore for small insects; larger insects may be dipped directly from the water. Many insects associated with water, such as dragonflies and damsel flies, will be available either in the juvenile stages in the water or as adults flying over the water.

In desert and semidesert situations where sagebrush and sand dunes prevail, both day and night collecting prove very profitable in season. One should use the sweep net or beat net on the standing brushy vegetation. Pay particular attention to any group of flowering plants, since these will be short-lived in the desert and will attract many insects. The moth sheet is an indispensable tool in night work for capturing a wide variety of insects. In the desert the collector will notice that populations shift, so that the daytime species often disappear in the late afternoon and are replaced by a crepuscular population (animals living in the gray hours), and that these in turn are replaced by the nocturnal segment which may appear in several distinct groups during the night.

KILLING, FIELD STORAGE,
AND PRESERVATION

Construction and Use of Poison Bottles

General Remarks.   Unfortunately, the most useful killing bottles are also the most poisonous and, hence, the most dangerous. Anyone working in this area must recognize that the cyanide poisons are deadly. It cannot be emphasized too strongly that parents or teachers take considerable risk in permitting young collectors to use such devices unless they are thoroughly reliable and trustworthy individuals.

Potassium Cyanide Bottle.   To make this killing bottle, select a strong glass bottle with a tight-fitting, one-piece lid equipped with a wax sealer or a strong, heavy-walled glass vial equipped with a rubber stopper. Unless large insects are being collected the bottle should be 5 or 6 inches tall and not much more than 2 inches in diameter, so that it will readily fit into the pocket. Put from ¼ to ½ inch of potassium cyanide in the bottom of the bottle and add ½ inch of sawdust over this (Fig. 12–5, A or B). After this is tamped in place, pour in about ⅜ inch of plaster of Paris which has been freshly mixed. If the bottle is tapped down on the table top the plaster of Paris will flow evenly to all parts of the bottle. Put the bottle in a hood or some other place where it can safely dry out. The sawdust will absorb some of the moisture from the plaster and in turn wet the cyanide. After the lid has been placed on the bottle, a day or two may be required before the cyanide permeates the plaster and "charges" the bottle. Crumpled paper or some dry leaves should be placed in the bottle for insects to hide in, and the word "poison" must be boldly printed across the outside of the bottle. The portion of the bottle from the plaster down to the bottom should be taped with adhesive tape to prevent the loss of cyanide in the event that the bottle is broken. Figure 12–5I shows how to dispose of a broken bottle in the field by digging a hole and placing rocks and dirt over the top of the bottle. One general drawback to the potassium cyanide bottle is that the bottle often sweats and becomes wet enough inside to damage and discolor some insects. Changes in temperature or humidity may start this sweating. Thus, the bottle must be moist enough to activate the cyanide and dry enough to prevent excessive sweating. The use of paper toweling in the bottle will help greatly.

Liquid Poisons.   Liquid poisons are generally safer than cyanide when used in killing bottles, but not as effective in killing insects. Ethyl acetate, carbon tetrachloride, or ether may be used. Select a stout bottle as described

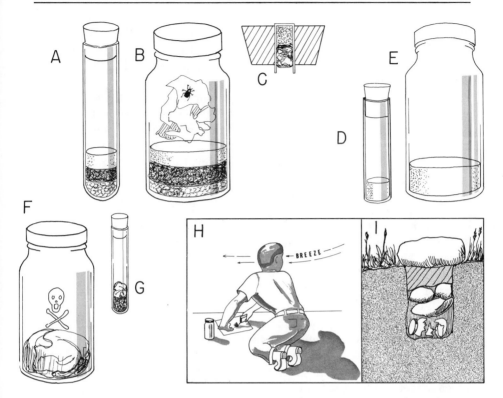

FIG. 12–5. Killing bottles. A–B. Potassium cyanide method. C. Cork stopper with cyanide vial. D–E. Ethyl acetate method. F–G. Cyan-o-gas method. H. Constructing a Cyan-o-gas bottle. I. Disposal of broken cyanide bottle.

above and pour into this ¾ inch of plaster of Paris (Fig. 12–5, D and E). After this has dried, saturate the plaster with the liquid poison. Crumpled paper or dried leaves should be added as a baffle for the insects. This will prevent them from fighting while they are in the process of dying. Specimens may be left for many days in a poison bottle made with ethyl acetate, provided the bottle is tightly sealed. If the ethyl acetate evaporates the specimens will become dried, and they must be relaxed before pinning. The liquid poison bottle is not too suitable for butterflies and moths as the wings become wet and thus are ruined.

Cyan-o-gas. Perhaps one of the best killing agents, for butterflies and moths in particular, and other insects in general, is Cyan-o-gas, a commercial ant poison available at most hardware or nursery stores. This comes in ¼-pound and 1-pound tins and is absolutely dry. Strong bottles or vials with tight sealing lids should be selected for this killing bottle as described above. A bottle of this type (Fig. 12–5F) should be made under a hood or out of doors where a breeze will dissipate the fumes (Fig. 12–5H). Spread three large paper table napkins on the ground, one on top of another. Pour approximately one rounded teaspoon of Cyan-o-gas on top of the first napkin and loosely roll this up in such a fashion that the chemical will not leak out. The first towel is rolled into the next and this into the third paper napkin so that a round ball of loosely crushed paper is formed. Force the finished ball, which is almost too large to go into the top of the bottle, down into the bottle and crush it into place with a blunt stick. This will hold the paper and Cyan-o-gas in place. Face the breeze while the bottle is being made so that the fumes will be blown away from you. When the lid is placed back on the bottle the Cyan-o-gas is ready to be used. A minor drawback in the Cyan-o-gas bottle is that specimens will be dehydrated somewhat more quickly than in other kinds of killing bottles and should not remain inside for long periods of time. Also, any cyanide preparation may affect colors of some bees and wasps if they are left exposed to it too long.

Small Cyan-o-gas vials may be made by simply pouring some Cyan-o-gas into the bottom of a vial and tamping enough paper toweling on top to keep the powder from escaping. A third technique that may be used is to fit a large cork (Fig. 12–5C) with a vial. The vial is filled with Cyan-o-gas and a paper toweling plug. The cork then is put back in the killing bottle which is then ready to use.

The Use of Poison Bottles.   It is unwise to mix robust insects, such as large beetles, with delicate insects, as the latter will almost always be partially destroyed. Butterflies and moths usually contribute so many scales to other insect specimens that mutual damage is done by mixing them together. Make it a habit to keep the killing bottle dry inside by changing the toweling frequently. Learn automatically to hold your breath while the top of the cyanide bottle is open.

Most insects can be tossed directly into the bottle while the lid is briefly removed. Bees, wasps, butterflies, and similar insects cannot be handled freely, and thus a special technique is needed for putting them in the bottle. In handling such insects, loosen the lid of the poison bottle and place the bottle on the ground. Place the net over the poison bottle and hold the tip up while the insect instinctively goes to the top of the bag. Take hold of the lid through the net and remove it while the specimen is worked into the bottle. Then replace the lid and screw it tight. Cyanide will alter the yellow color of bees and wasps left too long in the bottle.

Large butterflies are most conveniently demobilized by pinching the thorax through the butterfly net. This is done by allowing the wings to fold up over the thorax in a normal position and then grasping the butterfly through the net and firmly pinching the thorax without crushing. Then the butterfly is removed from the net and dropped into the killing bottle without wing damage.

Collecting Juveniles and Adults in Liquids

Many adult species may be collected into 70- to 80-percent isopropyl (rubbing alcohol) or ethyl alcohol when it is inconvenient to pin or layer in the field. Insects with furry or scaly bodies or with certain kinds of colors may be damaged in liquid preservatives. However, a wide variety of specimens can be placed in alcohol and pinned later with no damage whatever to the specimen. Juvenile specimens such as larvae and pupae are collected and maintained in liquids. If these are heavily chitinized and are to be studied externally they may be killed and preserved in alcohol. However, with soft-bodied specimens particular care should be taken in killing.

Specimens should be killed so that important taxonomic characteristics can be studied. Among these characteristics are various head structures, mouth parts, antennae, appendages, hooks, spines, bristles, hairs, spinnerets, and the like. Color may be of importance and thus color notes should be taken (see Chapter 1).

A number of methods and fluids for killing which will either harden or expand the specimen for study have been devised. If internal tissues or structures are to be studied, FAA or Bouin's fixative may be used (see Appendix C). Soft-bodied larvae such as mosquito wigglers, butterfly caterpillars, and so on, may be killed by dropping them in near-boiling water for a few moments and then transferring them to a preservative, usually 80-percent ethyl alcohol. Peterson (*Larvae of Insects*, 1959) recommends several killing solutions.

The Peterson XA solution will distend and harden soft-bodied butterfly, moth, beetle, and other larvae. Specimens should be removed after a day or so and transferred to alcohol. The XA solution consists of one part xylene and one part 95-percent ethyl alcohol.

Peterson's XAAD solution is used for beetle or butterfly larvae; it consists of four parts xylene, six parts isopropyl alcohol, five parts glacial acetic acid, and four parts dioxane. (*Note*, dioxane is poisonous.)

Peterson suggests that the KAAD solution is perhaps the best for killing most kinds of larvae. He cautions that bright colors may be lost, but nevertheless finds that this solution will distend and harden butterfly, fly, beetle, wasp, bee, and other larvae except for those with thick exoskeletons. This solution is made of one part kerosene, seven to nine parts 95-percent ethyl alcohol, one part glacial acetic acid, and one part of dioxane. Students interested in working with immature insects should see Peterson (1959*b*) and Chu (1949).

Field Storage of Insects

On longer field trips it is often necessary to store insect specimens temporarily so that they will not be damaged as a result of drying or mishandling before they can be properly prepared in the museum. Many different techniques must be used, depending on the kind of material that has been collected and the conditions under which the collecting has occurred. These will be discussed below.

Field Pinning and Pinning Boxes.   Unless the collector is going to be in the field for many weeks, so that the number of specimens taken is prohibitive, one of the best methods for dealing with specimens is to pin them into temporary field boxes, using mounting methods described below. Roaches, grasshoppers, earwigs, larger true bugs, flies, beetles, bees and wasps, moths, etc., are often handled in this manner. Serviceable field-pinning boxes may be made out of stationery boxes or similar cardboard boxes with tight-fitting lids, or of old cigar boxes. Fine-grain fiber wallboard (CeloteX) or polyethylene foam make excellent pinning surfaces. These should be cut to the inside dimensions of the box and securely fastened, either by gluing or by pushing pins in through the outside of the box. If styrofoam is used as a pinning base, paradichlorobenzene (PDB) should be avoided as a fumigant, as it will destroy the plastic. In its place moth crystals (naphthalene flakes) can safely be substituted.

After the insects are pinned they may be arranged in rows across the top of the box starting from left to right. A complete set of field data should be placed on the first pin of a series from a particular collecting station. This then will separate one series from the next so long as the insects are placed in rows. The alternative, of course, would be to put complete data or field numbers on each insect pin.

Field Storage in Liquids.   Any of the liquids described above may be used for temporary or, in most cases, permanent storage. Ethyl alcohol in concentrations of 75 to 80 percent is by far the best. One must be careful not to dilute the alcohol too much by placing too many insects into it. To avoid this danger the collector may change the alcohol on a full jar of specimens after a day or two has elapsed. Insects that are very hairy or covered with scales, such as bees and butterflies, should not be preserved in liquids. Alcohol and other liquids may affect insect colors drastically, and thus insects which have pale yellow or light green colors should not be placed in alcohol. Most insects with metallic colors or with black or dark brown colors will not be affected by alcohol.

Ethyl acetate serves as an excellent temporary storage medium for many insects, but insects whose color is readily affected by alcohol should not be stored in ethyl acetate. To store specimens such as beetles, equip vials with a small amount of paper toweling tamped firmly into the bottom. This, in turn, is wetted by ethyl acetate, the insects and field data are placed in the vial, and the vial is tightly capped or corked. Once sealed, the vials may be taped with plastic tape around the cap and stored in larger jars. This technique proves very useful for storing as long as six months or more, and will render specimens in excellent condition for genitalia dissections or routine pinning. The only danger with this technique is that the ethyl acetate may evaporate, leaving dried specimens, or it may pull grease from the bodies of insects which will necessitate a washing in ethyl acetate before pinning can be completed.

Papering.   Large winged insects such as dragonflies, damsel flies, butterflies, and the like may be stored in paper envelopes. These are folded from rectangular pieces of paper of suitable size, as illustrated in Fig. 12–6, A through E. One of the best methods for making envelopes is to purchase several sizes of small note pads of a good quality of white paper and to fold the envelopes from the pads as they are needed. The data should be recorded on each envelope before the specimen is placed inside. The data should include the name of the species or kind of insect if this is known. Bend the tips of the envelopes over (Fig. 12–6D) to ensure that the specimen will not slip out. Papered insects are then placed in any convenient box, along with a few moth crystals, and taken to the laboratory for pinning at any later date.

Layering.   One very common field method for storing large quantities of small insects is layering. Select a box with a tight-fitting lid for this technique. After a large number of specimens have been caught they are placed upon a layer of facial tissue or cello-cotton in the bottom of the box. Regular cotton or any material on which the insects will snag their legs should be avoided. Write the field data on a slip of paper and place this with each layer of insects. Additional facial tissue or cello-cotton is added and subsequent layers of insects and tissues may be placed in this box until the box is full. A few moth crystals should be added before the box is taped closed. In the laboratory the now-dried insects will have to be carefully removed with a forceps and placed in a humidity chamber for relaxing. It is possible to relax the entire contents of the box by placing it in a re-

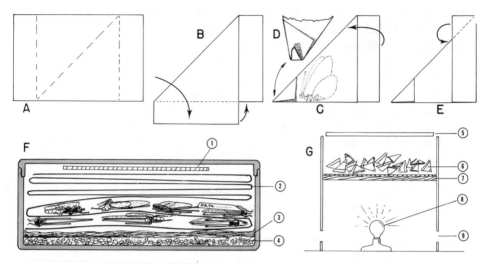

FIG. 12–6. Field storage techniques. A–E. Folding butterfly envelope. F. Sandwich box with chlorocresol and specimens. G. Field drying box. (1) Blotter with water, (2) folded toilet paper, (3) cello-cotton or tissue, (4) chlorocresol, (5) vent, (6) papered specimen, (7) shelf, (8) heat source, (9) air intake.

laxing jar, but the problem of molding and improper relaxation may develop.

Chlorocresol Method. A new method which removes the necessity of field pinning or papering all kinds of insects has been described by Tindale (1962) as the chlorocresol method. This method has tremendous advantages over field pinning and papering, not only in reducing the large amount of meticulous work necessary but in the fact that the specimens are already relaxed when they are taken out of the collecting boxes 6 months or more later. For this technique plastic sandwich boxes (Fig. 12–6F) are used, although other containers which are waterproof and can be sealed will be satisfactory. About a teaspoonful of chlorocresol crystals is placed in the bottom of the sandwich box. This, in turn, is covered by a tight-fitting layer of cello-cotton or other material which will keep the crystals from leaving the bottom of the box. Toilet tissue is then folded back and forth into the box to form an accordion, between the layers of which the insects are ultimately placed. A strip of blotter paper ½ inch by 3 inches is placed in the top of the box. In the field the specimens are placed between the layers of toilet tissue, beginning at the bottom and working toward the top. A sheet containing the field data should be placed in with each layer of specimens or at the top of the specimens from one particular locality. Between collections the boxes are sealed with masking tape to prevent loss of moisture. When the box is filled the blotter is moistened with water, placed inside, and the lid tightly taped in place. Upon returning from the field it is desirable to store the chlorocresol boxes in the refrigerator (not the freezer) until the insects are to be pinned. Upon opening the box you will find the specimens perfectly relaxed and usually needing no moisturizing before being pinned. Do not take too many specimens out of the container at one time, however, as some may dry out prior to pinning.

Chlorocresol functions essentially as an insecticide and fungicide and prevents molding and decay while the specimens are in storage. This product will become generally available on the market, although currently it is being imported from British pharmacy houses. It is available through Bio Metal Associates, Box 61, Santa Monica, California.

Field Drying in the Tropics.   In any damp, humid climate it is often essential to dry specimens so that molding will not occur, unless the chlorocresol method is employed. Butterflies that are papered or insects that are field-pinned or layered must be sufficiently dry to overcome destruction by moisture. Figure 12–6G shows a cross-section through a simple drying box which may be used in the drying of both plant and animal material. This may be constructed of a cardboard box which is provided with a screen shelf, air holes cut at the base to permit a steady flow of air, and a lid placed over the top to prevent rapid loss of heat. A heating unit such as an electric light bulb will send a draft of warm dry air up through the pinned or papered specimens until they are dehydrated. They are then sealed into containers with some moth crystals. If a gasoline lantern is used as a source of heat, the box must be set high enough so that the heat will not injure the specimens nor burn the box. A sheet may be draped around the box and down around the lantern to funnel the heat up through the specimens.

Field Data.   Specimens are of little value without adequate field data. One must recognize the danger of losing field data or failing to remember data that seemed so obvious at the time of collection. A field notebook should be kept in addition to the data that are placed with the specimens to provide a second source of information. Many workers simply put a field number with stored specimens, but they run the risk of losing the original data. The best method is to place a brief copy of the collecting data along with each group of insects preserved. This should include the locality and data along with the station number (as discussed in Chapter 1, p. 14). The field notebook should include in addition to the date and exact locality of collection, information concerning the vegetation associated with the specimens at the collection site, temperature and other weather conditions, and a description of the microhabitat. Using two sources of data in the field will ensure the value of the specimens that are brought back to the laboratory.

MOUNTING
INSECT SPECIMENS

The Construction of Mounting Equipment

In the preparation of insect specimens a few simple pieces of equipment are essential and others are greatly beneficial. Almost every common piece of equipment can be purchased through biological supply houses, but they can also be made at home at a fraction of the cost. Some of the more important pieces of equipment are discussed below.

Butterfly Boards.   The butterfly boards that are the easiest and cheapest to build are made from fiberboard, wallboard, or CeloteX. A very soft fine-textured board should be selected, rather than the coarse fibrous variety. Some companies make fine-grained wallboard with a glazed white paper surface which is excellent for pinning boards. Many sizes of pinning boards will ultimately be needed. Some of the common sizes are as follows: 3-inch width with a $\frac{1}{16}$-inch groove, 3-inch width with a $\frac{1}{8}$-inch groove, 4-inch width with a $\frac{1}{4}$-inch groove, 4-inch width with a $\frac{3}{8}$-inch groove, 6$\frac{5}{8}$-inch width with a $\frac{5}{8}$-inch groove. The length of the board is optional, but 16 inches proves to be very economical.

With a very fine-tooth blade (preferably on a table saw) the wallboard should be first cut to the length of the pinning board and then ripped to the desired width. Figure 12–7, A, B, and C, shows the more simple plans of construction. All of the pieces are ripped to the proper width and a groove of the appropriate size is ripped down the middle (Fig. 12–7A). A series of saw cuts rather than a dado blade should be used for making the groove. Two identical pieces are then glued and pinned together to form a single finished board. The advantage of this style is that the butterfly pin is thrust through two $\frac{1}{4}$-inch thicknesses of wallboard rather than a solid $\frac{1}{2}$-inch piece, as in Fig. 12–7B.

For the second design (Fig. 12–7B), rip a baseboard to the proper width and two pinning boards narrow enough to provide a groove when the three pieces are assembled. Wood glue and pins should also be used for assembling this board.

Many workers prefer to have tapered pinning boards so that the butter-

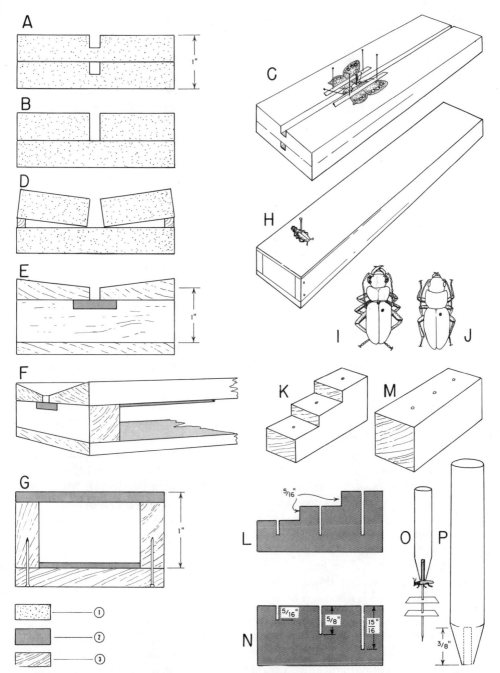

FIG. 12–7. Pinning boards and equipment. A–D. Butterfly boards, CeloteX construction. E–F. Conventional butterfly board. G–J. Insect mounting boards with specimens. K–L. Three-step pinning block. M–N. Three-hole block. O–P. Pinning dowel. (1) CeloteX, (2) balsa wood, or polyethylene foam, (3) soft wood.

fly wings will be slightly elevated. Even with proper drying butterfly wings sometimes snap down a little when the specimen is removed from the pinning board, or the wings may droop if the specimens are stored under moist conditions. Figure 12–7D shows a simple tapered pinning board made of strips of CeloteX that have been elevated with small wooden spacers (balsa wood or other). In the construction of this pinning board, adequate amounts of glue should be placed where wood surfaces contact, and pins should be used to secure the parts. The pins should not be driven from the pinning surface, but rather from the ends or sides of the pinning board.

Many workers prefer a medium-hard pinning surface made of some soft, fine-grained wood. Figure 12–7, E and F, shows the end and side views of a tapered pinning board made with wood. This model is constructed by first cutting a baseboard of the appropriate length and width and then nailing to it two end blocks. The end blocks should be thick enough so that the finished pinning board measures 1 inch from the pinning surface down to the top of the baseboard. The boards making up the pinning surface should be of a fine-grained soft wood and may be either tapered or flat. A ¼-inch thick piece of balsa wood should be fitted into a receiving groove in the end blocks and the pinning surfaces nailed or glued in place. In mounting butterflies, the pins are thrust through the balsa wood down to the top of the baseboard.

Insect Boards.   Insects should be mounted as neatly as possible; this often requires propping up the legs or body while the specimen dries (as discussed below) Fig. 12–7, I and J. To make this task easier collectors who are specializing in certain families of insects often use insect boards for the pinning and leg arrangement of specimens. Figure 12–7, G and H, shows the construction of this simple board. It consists of a baseboard, two side boards, and two end boards. These may be made of plywood or any suitable soft wood which is cut to the proper size, and assembled by means of nails and glue. A $\frac{1}{16}$-inch piece of balsa wood is placed in the bottom of the pinning board and a ⅛-inch piece of balsa wood is placed over the top as a pinning surface. The finished board should measure 1 inch from the top of the pinning surface down to the baseboard, as shown in Fig. 12–7G. The use of this board is discussed below.

Pinning Blocks.   Some form of pinning block is absolutely essential for placing the insect and its labels at the proper height on the pin. The pinning block is made of any fine-grained soft wood in the three-step pattern (Fig. 12–7K) or level pattern (Fig. 12–7M). The three-step pinning block is cut so that each step has a surface 1 inch wide and a rise of $\frac{5}{16}$ inch. After cutting and sanding the block put a $\frac{1}{16}$-inch metal drill in a drill press and set the gauge to drill $\frac{15}{16}$ inch from the top of the upper step. All three holes are drilled with this same setting, thus giving the required depth for each step (Fig. 12–7L).

The level pinning block is more easily constructed and is made of fine-grained soft wood. Three $\frac{1}{16}$-inch holes are drilled to the depths of $\frac{5}{16}$ inch, $\frac{5}{8}$ inch, and $^{15}\!/_{16}$ inch; a hand drill or drill press may be used for this operation. The holes are appropriately numbered 1, 2, and 3, to indicate the depth.

Pins and Pinning Forceps.   About ten sizes of insect pins, based on diameter, are usually available; they are scaled from fine diameter to large as 000, 00, 0, 1, 2, 3, 4, 5, 6, and 7. The first three sizes are good for direct pinning of very small insects, but they are difficult to thrust into a pinning bottom without the aid of a pinning forceps. Sizes 1, 2, and 3 are the most useful; 4 and 5 may be of some value with larger specimens. All of the sizes from 000 to 5 are approximately $1\frac{1}{2}$ inches in length. Size 6 is $1\frac{3}{4}$ inches and size 7 measures $2\frac{1}{16}$ inches in length. The larger sizes are sometimes useful for pinning giant tropical specimens. Pins may be obtained in the stainless steel or the black japanned finish. The japanned finish is less expensive and much more satisfactory than the stainless steel.

A pinning forceps should be available for transferring specimens into permanent storage cases. Special forceps are available, or a pair of long-nose pliers which has the jaws set at a 90° angle (this can be ordered through any hardware store) will serve just as well. Extremely lightweight forceps listed as pinning forceps are of little value. In using the forceps hold the pin right near the point and thrust it straight down into the pinning bottom. Be careful to line the pin up in the grooves on the jaws of the forceps, and not to snap the pin, for this will damage the specimen.

Insect Labels.   Insect labels are commonly printed by hand with a crow quill pen and waterproof India ink. The labels should be quite small (about $\frac{3}{8}$ by $\frac{1}{2}$ inch) so that they do not take up too much space and yet are large enough to protect the specimens. Labels may be used to give the collecting locality or the collector's name or even taxonomic data. It is not essential to outline the labels. If a large number of insects have been collected at one locality, hand printing is too time-consuming and tedious. Two convenient alternatives exist: the photographic reduction method and the printing press.

In the photographic method labels are typed on a white sheet of paper with a clean new ribbon. The individual labels are spaced in columns so that they can easily be cut with a paper cutter after reduction and are provided with room for the date of collection and the collector's name. The entire sheet is marked for reduction to 3 inches wide, and then turned over to a photographer. Figure 12–8A shows a part of a typed page at normal size and Fig. 12–8B shows the entire page reduced and printed. The photographer should have a camera with a ground-glass viewer so that he can measure the reduced size of the plate. An 8½- by 11-inch sheet of paper reduced to a width of 3 inches will produce the proper size of label. In this procedure no enlarging process is needed for printing the finished labels; rather, contact prints are ordered. Any time new labels are needed the negative is sent to the drugstore and contact prints made. One should request dull or mat finish on the prints, as printing with India ink is somewhat easier on such surfaces. Figure 12–8, C and D, shows other labels which may be made by this method.

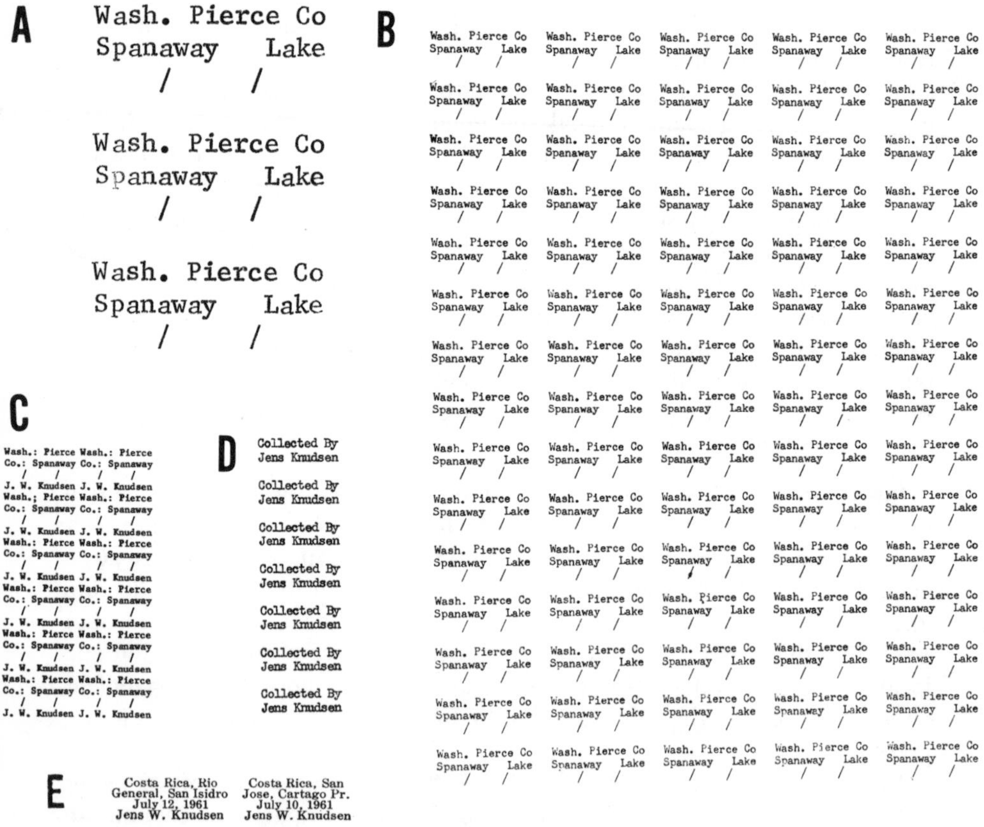

FIG. 12–8. Insect labels. A. Typed labels actual size. B. Photographically reduced labels finished size. C–D. Specimens of photographic labels. E. Printed labels of 4-point type.

Figure 12–8E gives an example of insect labels printed on a museum press. The labels shown are actual size and have been set in number 4 type. At the time of writing of this book, a small 3- by 5-inch hand press equipped with ink and a galley of type costs only about $50 (Kelsey Co., Meriden, Conn.). Four or five labels are set at one time and printed over and over again on 3- by 5-inch cards until the desired number is obtained.

Labels should always be turned on the pin so that they are read from the rear of the insect or the left side of the insect. One should be consistent in label position, but may use whichever method requires the least space.

Methods and Problems of Mounting Insects

Where to Pin Insects.   Insects must be pinned when they are fresh and pliable. Figure 12–9 shows where some of the more common insects should be pinned. Most insects are pinned eccentrically so that important taxonomic characteristics which may occur along the median line will not be destroyed. Beetles (Fig. 12–9A) are pinned through the right elytron, or wing cover, close enough to the front so that the pin passes between the second and third walking legs. True bugs (Hemiptera and Homoptera, Fig. 12–9B) are pinned eccentrically through the scutellum. Grasshoppers and their allies are pinned eccentrically either with one wing spread (Fig. 12–9C) or with both wings back in the normal position. The legs should be propped up and the antennae brought back along the sides of the body while the specimens are drying. Butterflies, moths, dragonflies, and their allies may be pinned (Fig. 12–9D) centrally and mounted on a butterfly board, as discussed below. Flies, bees, wasps, and their allies may be pinned eccentrically through the prothorax, as illustrated in Fig. 12–9, E, F, and G. Sometimes wasps are

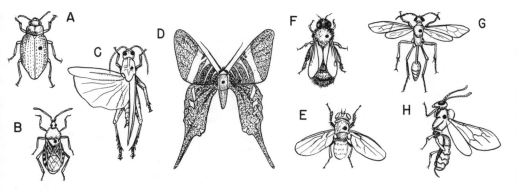

FIG. 12–9.   Pinning insects. See text for instructions.

TECHNIQUES FOR INSECT ORDERS

Order	Collecting Techniques for Adults	Killing Adults	Preserving Adults	Killing Juvenile Nymphs, Naiads Larvae, Pupae	Preserving Juvenile Nymphs, Naiads Larvae, Pupae
Protura	1,7	8	17,20,21	8	17,20,21
Collembola	1,7	8	17,21,21	8	17,20,21
Diplura	1,7	8	17,21	8	17,20,21
Thysanura	1,7	10,8	19	8,10	17,19
Orthoptera	1,3	10	18,23	10,11	18,19,17
Dermaptera	1,2	10,8,9	18	10,8,11	19,18,17
Plecoptera	1,3,5	10	18	8,9,11	17
Isoptera	1,2	9,10,11	19,18,17	9,10,11	19,17
Embioptera	1,2	10,8,9,11,14	18,17	8,9,11	17
Odonata	3	10	11,18	8,10	17,19
Ephemeroptera	3	8,10	17,18	8	17
Mallophaga	6,2	8,11	20	8	17
Anoplura	6,2	8,11	20	8	17
Corrodentia	2,7	8,9	17,20,21	8,9	17,21
Hemiptera	1,2,3,5	10,8	18,19	10,8,13	19,17
Homoptera	1,2,3,5	10,8	18,19	10,8,13	19,17
Thysanoptera	1,2	8	17,21	8,9	17
Neuroptera	1,3	10,8	19,17	8,9	17
Mecoptera	1,3	10,8	19,17	8,9	17
Trichoptera	1,3	10,8	19,17	8	17
Lepidoptera	3,4,5	10	18	8,12,13,16,15	17,22
Diptera	1,2,3,4	10,8,9	18,19,21	8,13,15,16	17,21
Siphonaptera	6,7	8	20,17	8,13,15	17
Coleoptera	1,2,3,4,5,7	10,8,9	18,19,17	8,12,13,15,16	17,19,18
Hymenoptera	3,1,4	10,8	18,19,17	8,11,13	17

Key

1. Hand collecting or with forceps.
2. Aspirator or paint brush.
3. Use an appropriate net.
4. Trapping or baiting.
5. Night lighting.
6. Capture and process host.
7. Berlese funnel, litter or nest material.
8. 75- to 80-percent ethyl alcohol.
9. 75- to 80-percent isopropyl alcohol.
10. Killing bottle (cyanide or other).
11. FAA for tissue studies.[a]
12. XA mixture.[a]
13. XAAD mixture.[a]

14. Ketone mixture.[a]
15. KAAD mixture.[a]
16. Boiling water.[a]
17. Store in 75- to 80-percent alcohol or other liquid (with or without glycerin).
18. Pin specimens.
19. Point on card.
20. Permanent slide (use KOH first).
21. Permanent slide (without KOH).
22. Inflate and dry.
23. Remove viscera of large fleshy specimens; pin. See text.

[a] Not suitable for all species.

pinned on their sides (Fig. 12–9H) when the abdomen is greatly curved, so that important taxonomic characteristics may be observed. The specimens should be propped up on a card until dried in this position. Most other orders of insects are pinned eccentrically or placed upon points or minute needles (*Minuten Nadeln*).

Use of Pinning Block.   Provide yourself with some suitable base upon which to start the process of pinning. This may be made by pressing clay into a bottlecap or by cutting a piece of ¼-inch balsa wood. Hold the specimen down between the thumb and first finger of the left hand and move your fingers back just far enough to expose the area that is to be pinned. Thrust the pin straight down through the body of the insect while firmly holding the specimen with the left hand (Fig. 12–10A). This technique will prevent the disarrangement of wings or other parts during pinning. Next thrust the pin down into the top hole of the three-step pinning block (Fig. 12–10B). Following this the locality label is placed on the second step of the pinning block; the collector's label or taxonomic label is placed on the third step. When these have been placed on the pin the specimen may be put away in a storage box. Figure 12–10, C and D, shows the correct way of pinning where the specimen is at the proper level on the pin and the body is level in both the longitudinal and transverse planes. Do not tilt the insect forward or backward or to one side or the other, and do not place it too high or too low on the pin, as shown in Fig. 12–10, E through H. If specimens have droopy bodies or long gangly legs (Fig. 12–10, I and J) they should be propped up on cards while the specimens dry. Drying may require a week or two for larger specimens. After drying, the card is carefully bent away from the insect and pulled down the pin, then the proper labels are added.

Relaxing Specimens.   Specimens that have been dried and stored, or specimens that have been in cyanide or Cyan-o-gas bottles too long, will be brittle and should be relaxed before pinning. Any jar or can that can be tightly sealed may serve as a relaxing chamber. This should be provided with some wet paper toweling, wet sand, or similar material, as a source of moisture, and with a shelf to support the specimens and keep them off the wet substrate. The shelf may be made of a piece of cardboard, screen wire, or a bottle lid. Specimens are stored in the relaxing chamber until the limbs are movable; then they are pinned. One must watch the relaxing chamber to prevent the growing of mold while the specimens are inside. A few drops of phenol or formaldehyde or a few chlorocresol crystals will retard molding. The size of the specimen and the degree of drying, plus other factors such as fat content, will determine the length of relaxation necessary.

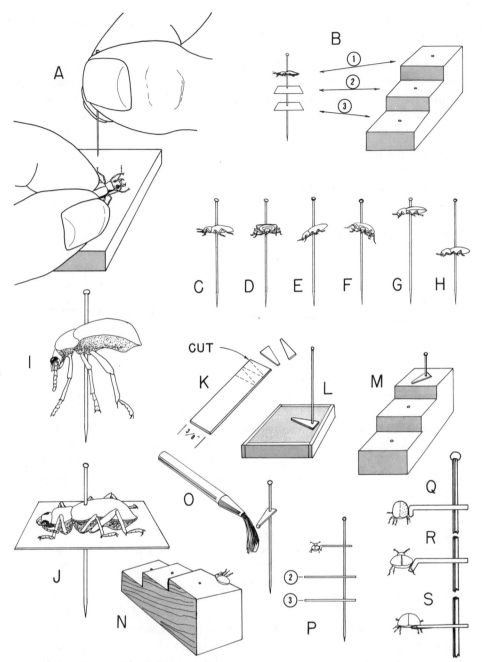

FIG. 12–10. Pinning techniques. A. Starting the pin with a pinning base. B. Use of pinning block. (1) Specimen level, (2) locality label, (3) collector's label. C–D. Correct method of pinning. E–H. Improper method of pinning. I–J. Propping drooping legs. K. Cutting points. L–M. Mounting point. N. Positioning specimen. O. Applying cement. P. Finished specimen with labels. Q–S. Bending point to match specimen.

Frequently, butterflies and moths will have to be relaxed from two or three days to as long as a week before the wings can be moved without damaging the specimen. Not all specimens relax at the same rate of speed; thus, some will be ready before others. Do not attempt to pin a partially relaxed specimen; rather, put it back until all of the body parts can be positioned easily.

Specimens may be relaxed quickly by using live steam. One method is to place a piece of screen wire over a ring stand, put the specimens on the screen, and cover them with an inverted tin can or beaker. Position the ring stand so that the screen is above the spout of a steam kettle (preferably of small size to reduce the volume of steam) and allow the live steam to penetrate the specimens until they are thoroughly relaxed. Only a few specimens at a time should be so treated. A second method is to place specimens in a shallow pan and then put a steam iron over the top so that live steam penetrates the bodies of the insects, relaxing them in just a few moments. Too much live steam may cause wing curling and distortion in butterflies.

Pointing Insects.   Insects that are too small to be pinned with regular insect pins should be mounted on insect points or *Minuten*. There is no rule which tells the collector when the insect is too small for a standard pin and should be placed upon an insect point. Of chief concern is the fact that a pin may destroy or damage a small insect. By definition, pointing consists of placing a triangular piece of paper on an insect pin and gluing the small insect to the tip. It is essential that insects be mounted in a uniform way with the point glued to the right side of the specimen's body (Fig. 12–10, P through S). Although many adhesives have been used in the past, by far the best is clear fingernail polish. This dries much faster than shellac, but slower than model airplane glue, and can readily be dissolved with fingernail polish remover, should the need arise.

Make insect points of a good grade of paper, such as manila folders or Bristol board, which will not fray after cutting. Cut points from strips of paper $\frac{3}{8}$ inch wide, as illustrated in Fig. 12–10K. The base of the triangle should be about $\frac{1}{8}$ inch wide and the tip $\frac{1}{16}$ inch or slightly less. The width and shape of the tip will vary, depending on the kinds of insects being mounted. Put the point on a number 2 or number 3 insect pin (never smaller), and position the point on the pin with a three-step pinning block (Fig. 12–10, L and M).

Next, position the insect on its left side on the pinning block so that the point can be applied to the right side of the body. If necessary, bend the tip of the point to conform to the insect's anatomy (Fig. 12–10, Q through S), add fingernail polish to the point (Fig. 12–10O) and attach the point to the side of the insect. Allow a few seconds for the fingernail polish to adhere to the insect's body and then lift the insect, pin and all, in a smooth easy motion and stick the pin into a pinning base. If the insect tends to roll or slip from the point you may correct this by pinning the specimen next to the three-step pinning block and resting the specimen on the step until the glue has dried. Using too much fingernail polish will only require a longer period of drying and partially obliterate important taxonomic features. Finally, place a label on the pin to complete the job. Do not neglect this important technique; it can easily be mastered with a little practice. Too frequently important smaller insects are omitted in collections because of their size. Most biological supply houses sell point punches which cut uniform points, and are well worth the price.

Minuten Nadeln (Minute Needles). *Minuten Nadeln* are pointed wire needles, about ½ inch in length, which are fine enough to pierce small insects and are attached to a larger insect pin by some suitable method. Figure 12–11, A and B, shows two methods of attaching *Minuten* to cork, balsa wood, or pith blocks. The technique in Fig. 12–11C is one in which the *Minuten* is mounted on a piece of file card attached to a larger insect pin. In all cases the pin should be of a number 2 or number 3 size and labels should be added as illustrated. Some collectors like to use special killing techniques so that specimens intended for *Minuten Nadeln* will be positioned naturally. For example, ether is often used for mosquitoes so that the wings and legs will be in a more or less normal position.

Carding Insects. This method may be used for smaller insects in place of pointing or *Minuten,* or for larger soft-bodied specimens that become very fragile when pinned and dried. The technique consists of cutting small squares of index card about the size of an insect label or of a size appropriate to the specimen to be mounted. The card is placed at the proper height on the insect pin and coated with a thin layer of clear fingernail polish. The specimen is then placed upon the card and becomes firmly

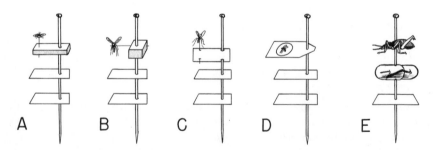

FIG. 12–11. Additional mounting techniques. A–C. Three methods of mounting minute insect pins. D. Coverslip mount. E. Using a medicine capsule for broken appendages.

attached as the cement dries. There are several drawbacks to this technique in that one side of the insect is hidden from view and overly large cards are space-consuming. When carding small specimens, mount them in different body positions so that all taxonomic characteristics can be viewed through the dissecting microscope. Perhaps this technique has its greatest value when very long series of small specimens are to be mounted for comparative studies, for dozens may be placed side by side on a single card.

Coverslip Mount.   When small insects are being prepared on permanent glass slides it is often desirable to place a few of these specimens in the pinned collection. In order to do this, prepare the insect in the standard way for slide mounting (discussed below and in Appendix B). A celluloid coverslip is cut (Fig. 12–11D) large enough to receive a 15-millimeter circular glass coverslip. The specimen is then transferred from the clearing medium onto the plastic coverslip as it would be onto a permanent glass slide. Add sufficient mounting medium to cover the specimen and lower the glass coverslip into place by means of a dissecting needle. Next, mount the plastic coverslip on a number 3 insect pin and attach the label. This technique will not replace the permanent glass slide, as specimens are difficult to study with a compound microscope when mounted by the coverslip method. However, specimens which might not otherwise have been included in pinned synoptic collections may be mounted by this method.

Mounting Winged Insects.   Special techniques must be used and care taken in the mounting of some winged insects such as dragonflies and damsel flies, butterflies, moths, and so on. Dragonflies and damsel flies are mounted on somewhat modified butterfly boards. These are constructed as described above (p. 150), but have a central groove measuring ¾ to 1 inch in width. This board not only allows room for the arrangement of the wings, but will permit the legs to be positioned as well. Wings are strapped down with paper strips as described below.

The mounting of butterflies and moths is a relatively simple procedure provided a few tricks of the trade are learned (see Fig. 12–12). The big difficulty, of course, arises from the fact that the bodies and wings are covered with scales which are easily dislodged and lost. Specimens must be fully relaxed before pinning.

After selecting the proper size of pinning board, carefully thrust a pin straight down through the center of the groove to provide a hole for the specimen pin; the pin used should be the same size or slightly larger than the one to be used for the specimen. This practice may prevent specimen damage that could otherwise result. Lift the specimen by the antennae and hold it by the legs, using the thumb and first finger of the left hand. The

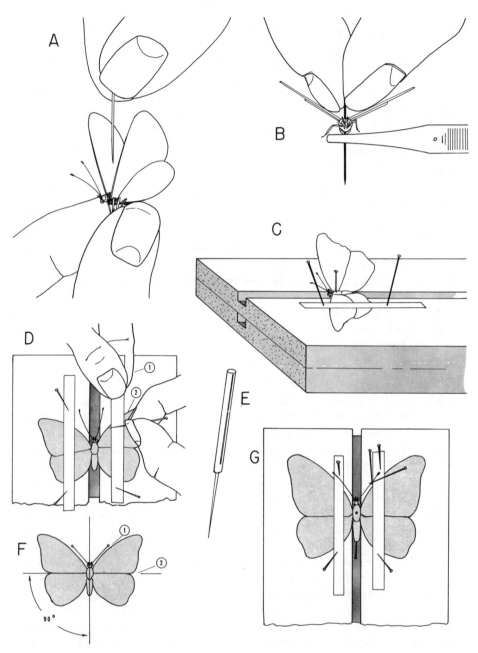

FIG. 12–12. Mounting butterflies and other winged insects. See text for instructions. (1) Proper spacing for antenna, (2) proper spacing of wings.

body itself should not be held; rather, it should rest firmly in the V formed between the thumb and first finger (Fig. 12–12A). Open the wings outward by blowing between them, maintaining a constant stream of air while the pin is lowered into position. If the wings are slightly stiff they may be opened first by inserting a closed forceps between them. When pressure is removed from the forceps the two blades will naturally open and cause the wings to separate. While blowing between the wings thrust the pin through the thorax. Be sure that the pin goes through the center and straight down so that the body is not tilted from side to side or from front to back. This is extremely important.

Next, grasp the head of the pin by the thumb and first finger so that when the specimen is pushed up on the pin, by forceps or fingers, the wings will strike upon the fingernails and thus will not lose their valuable scales (Fig. 12–12B). Always push the specimen higher up on the pin than is necessary. After the specimen pin is thrust into the mounting board, lower the specimen by means of forceps to the proper level. Care must be taken to have the wings exactly in line with the surface of the pinning board so that they will not be bent upward or downward when strapped in place.

Thread, string, glass, paper, and other materials have been used to strap down butterfly wings, but the author prefers to cut strips of slick-surfaced magazine covers and pages or 20-pound bond paper for this purpose. These should be 6 or 7 inches long and cut in ⅛-, ¼-, ⅜-, and ½-inch widths, as needed. Many workers like glassine paper because of its smooth transparent surface, but it is not very strong or durable. A strip of paper is inserted between the butterfly's wings and one set of wings is pinned down to the pinning board (Fig. 12–12C). Following this, the other set of wings is pinned down and then the wings are moved into the proper position (Fig. 12–12D). Move the wings with a fine 0 or 00 insect pin or a small probe made by thrusting a headless 00 pin into a matchstick (Fig. 12–12E).

The wings should be positioned so that the posterior margin of the front wing is at right angles to the midline of the body as shown in Fig. 12–12F (1 and 2). The hind wings are brought up in position so that only a small gap exists between them and the front wings. Pin the antennae at a uniform distance from the front wing, as shown in Fig. 12–12G. The abdomen should finally be propped up on a pin and carefully centered.

When moving the wings into position, pin the paper strap behind the wings and pull forward, with the left hand, to exert pressure upon the wing, as shown in Fig. 12–12D. Do not move the forewing all the way forward in one operation, as this often allows the hind wing to "jump" out and get above the forewing. While holding tension on the wing, (1) lift the paper strap slightly and (2) slide the forewing anteriorly by hooking a pin just behind the costa or radius vein. Move the hind wing up to meet the forewing, alternately exerting and releasing pressure with the paper strap, and continue moving the two wings in this manner until they are in position. With larger specimens it is sometimes necessary, temporarily or permanently, to hold the forewing by pinning it just behind the costa vein with a fine insect pin. The wings on both sides of the body are positioned in this manner and the paper straps repinned close to the margin of the wing. If it is desirable, glassine or plain paper may be pinned over the tips of the wings. The antennae and abdomen are adjusted at this time.

A great deal of damage is often done during removal of pins from the wings or removal of the specimen from the pinning board. Never pull the insect pins straight out of the wings or paper straps. Rather, twist the pin between your fingers until it is loose and then pull it free. First remove the pins from the antennae and the abdomen and then remove the paper straps. In the latter operation pull the strap forward and twist the forward pin loose, slide the strap out to the side until it clears the antenna, and remove the strap. Keep labels low down on the shank of the pin to make them visible from above.

Insect Boards.   Insect boards should be used when the greatest degree of pinning perfection is desired or when large numbers of "droopy" insects are to be pinned. The boards (described above) have a pinning surface of ⅛-inch balsa wood. After thrusting the pin through the insect's body, push the pin all the way down into the insect board (Fig. 12–7H), and position the appendages with a pair of fine forceps or a dissecting needle (Fig. 12–7, I and J). Pins required to hold the legs are thrust into the balsa wood and left until the specimen is dry.

Riker Mounts.   Riker mounts are very popular for displaying all kinds of plant and animal material, as well as insects. The Riker mount consists essentially of a cotton-filled, glass-topped box in which specimens are placed. These mounts are excellent for teaching displays and limited synoptic insect collections, but are quite unsatisfactory for growing taxonomic collections. The Riker mount will give good protection to specimens that are to be handled by students, but once they are sealed it is a lot of work to reopen them in order to add or rearrange specimens.

Small Riker mounts for one or two specimens are easily made (Fig. 12–13A) by cutting strips of wood ¼, ½, ¾ inch square. These are then glued or tacked around the margins of a piece of cardboard of appropriate size. This forms a simple box into which cotton and the desired specimens are placed. A piece of glass, cut to fit the top of the box, is taped in place with black photographer's masking tape.

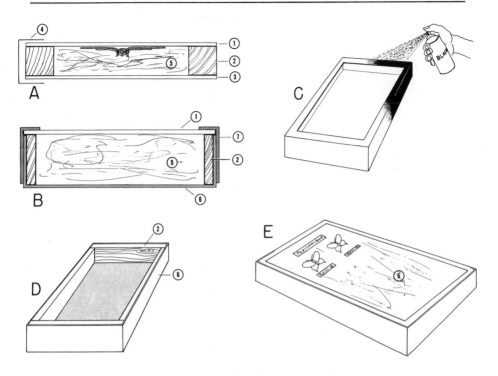

FIG. 12–13. Riker mounts. A. Simple construction. B. Box construction. C. The lid. D. Bottom with wood strips. (1) Glass, (2) wood strip, (3) cardboard back, (4) black tape, (5) cotton, (6) box bottom, (7) box top. E. Placing specimens on cotton.

Excellent Riker mounts can be made if boxes of the appropriate size, about 1 inch deep, are available (e.g., ditto stencil boxes). Remove the top of the box and draw a line ½ inch in on all sides. Place the top over a wooden block and cut along the lines with a scalpel, as shown in Fig. 12–13B. Spray the cover with a dull-finish black enamel paint. Next, cut some ¼- by 1-inch wooden strips to fit inside the bottom of the box (Fig. 12–13D), and glue these in place. Place a thick smooth layer of cotton in the box, followed by appropriate specimens and specimen labels (Fig. 12–13E). Finally, place a sheet of glass over the specimens and put the painted lid back on the box. The lid may be held in place with straight pins that have been cut off and thrust into the box or by black masking tape.

When Riker mounts are used for insect display a special technique should be used in preparing the specimens. It is essential that the specimens (butterflies, bees, dragonflies, and so on) have their wings and legs properly arranged before they are placed in the Riker mount. Sometimes specimens have a good natural position immediately after coming out of a killing bottle, but usually this is not the case. Therefore, specimens should be mounted in the conventional way until the wings and legs are set in the desired position. This poses a problem, however, since the body juices of fresh insects will weld the specimen to the insect pin. You may overcome this problem by papering butterflies and layering insects until they are dry. Then relax them and mount them in the conventional manner. After this has been done the insect is slid down the pin and placed in the Riker mount.

Microscope Slide Mounts.   On many occasions permanent slides are required in entomology. For example, studies of wings, mouth parts, entire head structures, genitalia, or small adult or larval insects such as fleas, lice, flies, mosquitoes, aphids, and the like, may all require slide-making technique.

Several kinds of mounting media are commonly used in entomology and should be selected for the material to be mounted. Complete instructions for the media mentioned below will be found in Appendix B or Chapter 13. Balsam, Permount, and Euparol require complete dehydration of the specimens prior to mounting. Other media such as Hoyer's, Turtox CMC-10, PVA, lactophenol, glycerin jelly, and others may be used with specimens taken directly from water, alcohol, or other preservatives without intermediate dehydration. The lactophenol medium, the Turtox medium, and Hoyer's medium have added advantage in that they will quickly kill larvae right on the slide and thus do the job of killing, fixing, clearing, and mounting. Some specimens will have to be macerated in KOH to render the exoskeletons clear. Stain usually is not required unless specimens are small and transparent and have a refractive index similar to that of the mounting medium. When thick specimens are mounted the coverslip may be propped up with broken pieces of glass or the specimens may be placed in a depression slide. Thick mounts usually require the addition of more mounting medium around the coverslip where the evaporation of solvents has started to form air pockets. Some instructions to be used in conjunction with Appendix B are mentioned below.

1. Whole mounts.   Fleas, lice, darkly pigmented insects, or large structures such as biting-fly heads and the like are first macerated in KOH and then mounted in balsam or Euparol. Small flies, aphids, and other small insects are not usually macerated in KOH but are cleared in clove oil or some other medium prior to mounting in balsam or Euparol.

2. Mouth parts and genitalia.   The head or abdomen of the specimen is removed and macerated in KOH. When working with small insects macerate the entire animal. Following this, complete the procedure required for the mounting medium selected (balsam, Euparol, Hoyer's, Turtox, glycerin jelly; Appendix B or Chapter 13) and transfer the head or abdomen to a slide with a few drops of mounting medium. Dissect out the mouth parts or genitalia on the slide and throw away the excess parts; add a few more drops of mounting medium and a coverslip. Glycerin jelly mounts are often desirable in that the specimens remain quite pliable during the period of dissection. Genitalia are often put in $\frac{1}{8}$-dram vials of glycerin, and pinned with the insect.

3. Wings and legs.   Dried wings or legs may be wetted in 95-percent alcohol and mounted in Euparol or Permount. Unless the elytra of beetles or large, dark, and opaque legs of insects are being mounted, no maceration in KOH is necessary.

4. Small larvae and eggs.   Small larvae usually are not macerated in KOH nor stained unless they are perfectly transparent. They may be killed on the slide and mounted in Hoyer's medium, lactophenol, or Turtox CMC-10 (according to Peterson, 1959, the Turtox medium does not kill quickly). Specimens to be mounted in glycerin jelly, balsam, or Permount should be killed in hot water (see p. 145) and then mounted according to Appendix B.

5. Dry mounts.   Bristles, hairs, scales, and similar structures may be too transparent for mounting in a standard medium and may reject stain; they should be mounted dry. In mounting, these structures are transferred to a slide and covered with a dry, clean coverslip. A wringing medium such as asphaltum or clear fingernail polish is used to seal the coverslip in place.

MUSEUM
AND STORAGE TECHNIQUES

Construction and Use of Equipment

Storage Cases.   Insect specimens must be stored in well-constructed cases with tight-fitting lids in order to protect them against insect pests and dust. Biological supply houses sell a wide range of insect cases which differ in size, quality, and price. However, insect cases of good quality may be constructed in the home or school workshop. The size of the case is optional, but should be small enough for convenient storage. If glass is being used, the case dimensions should be based on stock glass sizes to avoid the necessity of cutting and fitting glass. Glass is sized in graduations of 2 inches. For example, a common insect case sold commercially measures 17 by 19 inches, using a glass size of 16 by 18 inches.

When selecting materials for case construction, use a fine-grained soft wood, free of knots, for the sides. If plywood is used for tops or bottoms, $\frac{3}{16}$-inch or $\frac{1}{8}$-inch mahogany plywood is suitable and least expensive. A fine-grained fiber wallboard (such as CeloteX) makes an excellent pinning bottom although balsa wood, cork, or special bulletin-board-cork linoleum are very satisfactory.

In planning insect cases, size is the first consideration. Extra-large sizes should be avoided, and the size of storage cabinet in which the cases will be housed must be considered. The style of construction depends somewhat on the means of storage. For example, glass-topped boxes are excellent for permanent storage in light-proof cabinets. On the other hand, a solid light-proof construction must be used if storage cabinets are not used. Figure 12–14 shows several plans for constructing glass-top and solid cases. Note that Fig. 12–14, A, B, and D show three styles for the glass-top construction. Of these, Fig. 12–14A proves most satisfactory. Figure 12–14B shows the glass set into a wooden lid and requires extra craftsmanship. Figure 12–14, D and J, shows the construction of a sliding glass lid which is somewhat less satisfactory than that shown in Fig. 12–14A. Plans for the bottom construction are given in Fig. 12–14, A, B, and C, the latter being preferable. The bottom shown in Fig. 12–14A provides a better finish with the addition of plywood, but is more expensive; that shown in Fig. 12–14B is subject to chipping and abrasion.

When constructing a case like that in Fig. 12–14C, first select the glass size and then cut the $\frac{1}{2}$-inch CeloteX bottom to the same dimensions as the glass. In figuring the height of the side, plan on no less than $1\frac{1}{2}$ inches for the inside of the box, plus $\frac{1}{2}$ inch for the CeloteX and $\frac{3}{8}$ inch for the glass recess. If the inside dimension of the box is to be 2 inches deep, for example, the side boards should be ripped at $2\frac{7}{8}$ inches. Next, set the table saw for $\frac{3}{8}$ inch and cut out the recess for the glass. Do the same thing for the CeloteX bottom. Figure 12–14G shows the typical mitered joint of a corner (which is difficult to cut and nail). Preferably, construct the corner like that of Fig. 12–14, H and I. When cutting the side boards to fit the glass, remember it is the outside of the glass recess, not the outside of the box itself, that must be measured. Always add about $\frac{1}{16}$ inch extra to the size of the glass to prevent it from sticking. Glue and nail all corners in assembling the sides. Use finish nails which can be set and puttied. Nail the CeloteX in place with headed nails before the glue used in the side joints has dried. The bottom CeloteX will square the box and ensure a good fit for the glass. Finally, sand the entire box, apply dull black paint or varnish to the outside of the box, and paste white paper or glazed cardboard on the inner sides of the box. Set the glass in place, using a small piece of tape at one corner so that it may easily be lifted out again. When

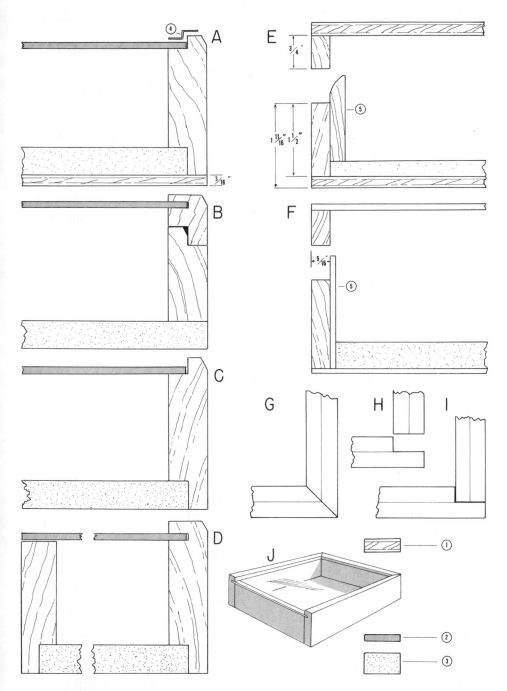

FIG. 12–14. Pinning cases. A–D. Top and bottom patterns for glass-topped cases. E. Solid case, plywood construction. F. Solid case, cardboard construction. G–I. Corner construction. J. Finished case as shown in D. (1) Plywood, (2) glass, (3) CeloteX, (4) black tape, (5) inner lip of box.

the case is filled with insects and is to be permanently stored, the glass may be taped in place with black photographer's masking tape (Fig. 12–14A).

Figure 12–14, E and F, shows cross-sectional views of the construction of solid-topped insect cases made of plywood or ⅛-inch cardboard. In both cases the lid fits tightly over an inside shoulder, resulting in an airtight insect-proof case. When using the plywood construction, first cut the top and bottom sheets of plywood to the exact finish and size of the box. Using ⁵⁄₁₆-inch clear, fine-grained wood, cut side boards 2⅜ inches wide (1½ inches bottom, ¾ inch top, plus ⅛ inch for a saw cut). Cut the sides and ends to the proper length, making a square or mitered corner. Glue and nails the sides together and then glue and nail the top and bottom onto the sides, thus making a solid box. When this is dry, set up the ripping gauge on the table saw and cut the top off from the bottom of the box. In doing this, set the saw blade so that it will cut only ⁵⁄₁₆ inch. Next, cut some suitable pinning bottom such as ⅜-inch CeloteX or ¼-inch cork and, if necessary, cover this with white paper. Place the pinning bottom into the box and then glue in the mitered inner shoulders as diagrammed. Carefully remove all excess glue and place the lid on the bottom of the box while the shoulders set and dry. This will ensure a tight fit. Finish the box with light sanding and a coat of shellac and wax or a coat or two of varnish.

Figure 12–14F shows a solid insect case constructed with ⅛-inch cardboard. The cardboard and side boards are cut as described above. The side boards are glued and nailed and the top and bottom cardboard are glued into place. When they are dry, saw the lid and bottom apart, add a shoulder piece of cardboard, and finally press a tight-fitting pinning bottom of ⅜- or ½-inch CeloteX into the bottom of the box. The pinning bottom is designed to hold the shoulder boards in place, and thus glue should be added to give additional strength to the box. This box may be unfinished or varnished or painted, as desired.

Insect-Case Layout.   One of the most frustrating jobs in entomology is to determine a policy for collecting, for this will govern the way the insect cases should be laid out. To begin with, a taxonomic layout is always used where insects are grouped by orders, families, genera, and species. When adequate literature is not available, beginning collectors key their specimens down to the family level and then group all members of that family under one heading. The problem that develops is knowing beforehand how much space to allow for each family or genus or species, for the job of moving thousands of pinned insects to make room for an additional group is tremendous.

Some of the decisions that must be made ahead of time are as follows:

it must be decided whether specimens will be identified beyond the order, family, or genus level; it must also be decided whether a general representational collection of each family, genus, or species is to be obtained or whether, ultimately, every species of a group will be sought. Also, one should decide whether he will collect and pin one specimen of each species or up to several hundred of each species. This will depend on the use of the specimens, whether they are for display and teaching or for taxonomic statistics. Some consideration in allotting space must also be given to the different sizes of insects. Many beginning collectors attempt to set aside room for orders and families depending upon their relative local numbers and general size. When insect cases with solid pinning bottoms are used, preplanning of space allotment is essential. This becomes much less serious, however, when unit pinning trays are used.

When using solid pinning bottoms, rule the bottom of the pinning case into columns 2 to 3 inches or more wide (the columns should be some fraction of the inside dimension of the box), as shown in Fig. 12–15C. Next, print or type labels for orders and families and affix these with ⅜-inch straight pins (sold as sequin pins in dime stores). These labels may be plain or ruled around the margins, according to preference (Fig. 12–15D). All of the insects belonging to the particular group, as labeled, are pinned in place. Leave room for additional specimens, then attach the next group label along with its insects, and so on.

Unit Pinning Trays.   Almost all large museums and serious collectors have adopted the use of unit pinning trays as opposed to the solid pinning bottom. Unit pinning trays consist of cardboard boxes of uniform width and depth, but varying in length. The sizes are designed to receive any group of insects, small or large. Usually one tray is reserved for each species, genus, or family, depending on policy. By this method, trays of insects can be rearranged and moved from case to case without one's having to repin the specimens. Figure 12–15A shows an insect case with pinning trays of various sizes inside. Each tray is provided with its own pinning bottom. Figure 12–15B gives some of the common dimensions for the pinning trays. Only the lengths are given, since the depth of the box is always 1⅝ inches and the width is usually 4⅜ inches (the width is a fraction of the insect-case width).

Unless a very large quantity of pinning trays is to be used, it is perhaps cheaper to buy these readymade (see Wards, Monterey, Calif.; Bio Metal Associates, Santa Monica, Calif.; etc.). Museums that make their own pinning trays order the boxes readymade at 1 or 2 cents each, and glue in a

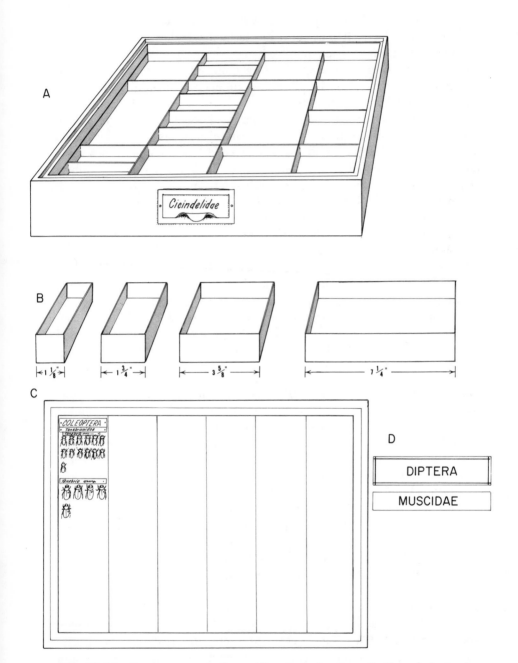

FIG. 12–15. Pinning cases. A. Case with unit pinning trays. B. Sizes of pinning trays. C. Layout of case with solid pinning bottom. D. Case labels.

piece of ⅜-inch CeloteX. One can make his own boxes by cutting cardboard to form the bottom and sides, folding the sides upward and taping the corners. This is tedious work and uniformity is often lacking in the finished products. (However, see p. 126 for technique.)

Liquid Storage.   Museums that keep collections of larvae, nymphs, or adults for study and dissection find it necessary to store specimens in liquid preservatives. Perhaps the best method is to keep specimens with their data in vials of preservative, stoppered with cotton (or a cork or screw cap). These, in turn, should be kept in wide-mouth museum bottles, along with enough additional preservative to cover the vials. A label placed in the bottle denotes the vial contents at a glance. This method reduces evaporation from the vials and requires little maintenance.

Special Storage Problems

Pests and Fumigation.   So-called "clothes moths" and dermestid beetles will quickly render hundreds of hours of work useless by destroying insect specimens. These pests must be kept out of insect cases by means of tight-fitting lids, tight storage cabinets, and fumigation. Fumigation should be a constant process in any museum, rather than one used only when an infestation has started. Paradichlorobenzene (PDB) or naphthalene flakes (moth crystals) should be added to each pinning case in small quantities about twice each year. The fumigant may be placed directly on the bottom of the case, but may leave a dust residue. Large pieces of PDB may roll around and damage delicate insect specimens. Workers usually make a small cardboard box for the fumigant and pin this in one corner of the case. PDB should not be used with a styrofoam pinning bottom, as PDB will make the styrofoam evaporate; use naphthalene flakes instead.

When looking for infections of dermestid beetles or clothes moths, look for what appear to be little piles of sawdust under the insect specimens. These may be on the labels or on the floor of the pinning box. If you observe them, remove the specimen and look carefully for a small hole where the larval pest has entered the specimen. Destroy the entire specimen unless it is valuable, in which case it may be especially fumigated with carbon bisulfide, strong PDB, or some killing agent. Next, fumigate the entire case of apparently uninfected insects with a large quantity of PDB or carbon bisulfide. Carbon bisulfide is very dangerous, and one should follow the precautions given on the container. Children should not be allowed to use or have access to these fumigants.

Light and Color.   Insect colors are produced either by pigments or by structural devices. Some common pigments are carotene, which produces yellow and orange colors; melanin, which produces light tan to black; metabolic waste products (uric acid derivatives), which produce white, light gray, and metallic silver; and plant pigments that have been ingested and deposited under the cuticle. Structural colorations are produced over some common pigment, such as melanin, by irregularities on the cuticle of the insect which refract light to produce the brilliant metallic colors. Structural colors are the most stable, colors deposited within the cuticle are next; pigments deposited just beneath the cuticle are likely to be altered by organic changes that occur after the insect dies. Light will affect any of these colors in time, some more rapidly than others. Change of color is extremely slow and hard to detect and damage may occur before one is aware of it. The only sound advice is to keep insects in a light-proof case when they are not in use. Short periods of exposure will do no harm, but longer exposures of many weeks or years should be avoided.

Dust and Cleaning.   Dust and lint will rapidly collect on insects that are not covered and will greatly alter their appearance and usefulness. Specimens that are greasy will have to be cleaned, as described below; those which are dry, but dusty, may be cleaned by gently brushing them with a water-color brush and then blowing the dust away with a fine stream of air.

Greasing and Degreasing.   Some insects, especially beetles, flies, bees, and the like, contain large quantities of stored fat which may in time work through the cuticle and discolor the specimens, both by saturating natural pigments and collecting dust and lint. Wings of butterflies may become so greased that the scales will be matted and discolored. When this occurs curators usually degrease specimens by placing them in a solvent such as carbon tetrachloride, ethyl acetate, or others. Probably the most effective solvent is clean xylene or xylol. Occasionally, some of these degreasing compounds will affect insect labels, but usually the specimen, label and all, can be pinned to the cork of a large vial and then placed in a vial filled with the degreasing solution. Degreasing may take from one to several days, but is well worth the effort.

Moisture, Drooping, and Mold.   Excessive moisture will penetrate the joints of appendages and eventually cause legs and wings to droop. With prolonged exposure to moisture, mold may begin to grow on certain insects which not only destroys them but ensures that spores will be distributed throughout the entire pinning case. The author once lost a box of extremely valuable butterfly specimens which were left by mistake in a damp basement in southern California. The storage room should be heated sufficiently to reduce excessive moisture.

Packing Insects for Shipment

Dried insects are shipped as pinned, layered, papered, or chlorocresol specimens, as described above. When shipping pinned specimens, outfit a strong wooden or similar box with a pinning bottom, such as polyethylene foam; pin the specimens deeply and securely into this. Cut a piece of cardboard just large enough to fit over the pins inside the box when the sides and ends of the cardboard have been folded upward. The sides and ends should be flush with the top of the cigar box, so that the box lid will hold the cardboard firmly in place. Tape all cracks and openings to the box and label the box as to its contents. Next, pack the specimen box inside a much larger pasteboard box and surround it on all sides with several inches of crumpled newspaper. Finally, tie and tape the pasteboard box (use paper gummed tape) and clearly label it on all sides with "Fragile" stickers. When declaring specimens for custom purposes, list them as museum specimens of no commercial value. If the box is to be shipped or hand-carried across a state or national border where agricultural checks are routine, the inclusion of paradichlorobenzene in copious quantities in the outer box seems to soothe the nerves of agricultural agents and also ensures against the transportation of some pest organism.

Living material is more difficult to ship in that problems of high or low temperatures, dehydration, feeding, agricultural regulations and laws, and the like, all have to be dealt with. Many states and countries have laws forbidding the shipment of living material and require that a prior permit be obtained. Air freight is the most suitable commercial method of transportation for live specimens. The container must be appropriately marked as having living material for research purposes and warnings to keep the specimens from extreme heat or cold.

REFERENCES

Belkin, John N., 1954, *Laboratory Manual for Medical Entomology,* printed by Dept. of Entomology, Univ. Calif. at Los Angeles.

Chu, H. F., 1949, *How to Know the Immature Insects,* Brown, Dubuque, Iowa.

Comstock, J. H., 1940, *Introduction to Entomology,* Comstock, Ithaca, N.Y.

Davies, D. A. L., 1954, On the Preservation of Insects by Drying in Vacuo at Low Temperature, *The Entomologist* 87:34–35.

Ford, E. B., 1955, *Moths,* Collins, New York.

Gaul, A., 1953, *The Wonderful World of Insects,* Holt, Rinehart & Winston, New York.

Harman, I., 1950, *Collecting Butterflies and Moths*, John de Graff, New York.

Harrison, A. S., and R. L. Usinger, 1934, Methods and Techniques, *Bull. Brooklyn Entomol. Soc.* 39:168–170.

Holbrook, J. E. R., 1927, Apparatus and Method Used to Remove Pins from Insect Specimens, *J. Econ. Entomol.* 20:642–643.

Holland, W. J., 1968, *The Moth Book*, Dover, New York.

Holland, W. J., 1949, *The Butterfly Book*, Doubleday, Garden City, N.Y.

Hussey, L., and C. Pessino, 1953, *Collecting Cocoons*, Crowell, New York.

Imms, A. D., 1951, *Insect Natural History*, McGraw-Hill, Blakiston Division, New York.

Jaques, H. E., 1947, *How to Know the Insects*, Brown, Dubuque, Iowa.

Klots, A., 1951, *Field Guide to the Butterflies*, Houghton Mifflin, Boston.

Knudsen, J. W., 1966, *Biological Techniques: Collecting, Preserving, and Illustrating Plants and Animals*, Harper & Row, New York.

Lutz, Frank, 1948, *Field Book of Insects*, Putnam, New York.

Merryman, H. T., 1961, The Preparation of Biological Museum Specimens by Freeze-Drying: II. Instrumentation, *Curator* 4(2):153–174.

Needham, J., and M. Westfall, 1954, *A Manual of Dragonflies of North America*, Univ. of California Press, Berkeley, Calif.

Peterson, Alvah, 1959a, *Entomological Techniques*, Edwards, Ann Arbor, Mich.

Peterson, Alvah, 1959b, *Larvae of Insects*, Edwards, Ann Arbor, Mich.

Ross, Edward S., 1953, *Insects Close Up*, Univ. of Calif. Press, Berkeley, Calif.

Ross, H., 1965, *A Textbook of Entomology*, Wiley, New York.

Swain, Ralph, 1948, *The Insect Guide*, Doubleday, Garden City, N.Y.

Tindale, Norman B., 1962, The Chlorocresol Method for Field Collecting, *J. Lepidop. Soc.* 15(3):195–197.

13

The Crustaceans

The crustaceans make up a class of arthropods (animals bearing jointed legs) which includes the so-called water fleas, fairy shrimp, barnacles, sow bugs, crabs, shrimp, and others. With the exceptions of some terrestrial and semiterrestrial crabs and isopods, etc., the crustaceans are aquatic and marine animals. They tend to be the "insects" of the aquatic world and occupy, to a large degree, all of the niches occupied on land by true insects. Because of the exoskeleton, the jointed appendages, and the well-developed sensory system, members of the class Crustacea are capable of a great degree of activity, may be fast-moving, and will require special techniques for capture.

REMARKS CONCERNING PRESERVATIVES

The only practical method for maintaining crustaceans is in liquid preservatives such as alcohol or formalin. (Slide mounts and other techniques may be used for limited numbers of specimens.) However, three problems must be considered: (1) the stiffening of muscle tissues in jointed appendages as a result of preservation, (2) the loss of pigments in larger crustaceans, (3) the destruction of those exoskeletons containing calcium carbonate. A solution of 70-percent alcohol, properly used, is the best general preservative in that jointed appendages do not stiffen and the solution has no effect on calcium carbonate. It does, on the other hand, quickly leach out the pigments of crabs and other such animals. If neutral formalin is used (see Appendix C) color retention is much better, although a disagreeable odor and the stiffening of muscle tissue are results.

REHYDRATION
OF DRIED SPECIMENS

Occasionally, museum jars are improperly sealed, causing the preservative to evaporate totally. The resulting dried specimens are fragile and do not respond well to represervation. However, if specimens are placed in a ½-percent solution of trisodium phosphate they will become thoroughly moistened and relaxed so as to permit the movement of appendages. After such a treatment they may be washed briefly in fresh water and returned to their original preservative.

SMALL CRUSTACEANS

General Collecting Methods

Processing Protective Plants.   Many small crustaceans hide among freshwater and marine plants and other debris, some even taking on characteristic colors of green, red, or brown to match the plant host. Coral blocks, branching corals, and hydroid colonies also harbor numerous crustaceans.

The simplest procedure for processing crustaceans is to suspend quantities of such weed in a bucket of habitat water with a very small quantity of formalin added. The formalin is irritating and, sooner or later, causes the animal to abandon the host plant, swim about in the water for a brief time, sink to the bottom, and die. The plants may be shaken in the water once or twice before removal to dislodge any additional specimens.

Birge Cone-Net Technique.   The Birge cone net is designed to work through vegetation of either shallow or deep water, and collects small organisms without including large quantities of weeds. The net (Fig. 13–1) is either towed behind a skiff or thrown out into weedy areas and retrieved.

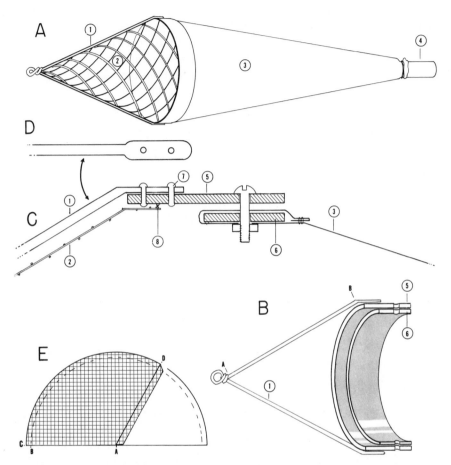

FIG. 13–1. Birge cone net. A. The assembled net. B. A cutaway diagram of the net rings. C. A cross-sectional diagram through the net. (1) Wire bridle, (2) screen wire cone, (3) cloth net, (4) receiving vial, (5) outer ring for wire cone, (6) inner ring for cloth net, (7) rivet, (8) solder screen at this point. D. A detailed drawing of the wire leader. E. Cutting the screen for the cone.

A simple, modified Birge net consists of a forward wire-screen cone, a muslin net, and a collecting bottle at the tip of the net (Fig. 13–1A). To construct this net, make a hoop of copper, brass, or galvanized iron ⅛ inch by 2 inches and about 8 to 10 inches in diameter, as desired. The metal should be overlapped and riveted or soldered to form the hoop. Next, make a smaller hoop of ¹⁄₁₆-inch by 1-inch metal which is approximately ⅛ inch smaller in diameter than the larger hoop, as seen in the cutaway diagrams

(Fig. 13–1, B and C). Make a towing wire from 9-gauge galvanized wire, twist to form a forward loop, and bend to fit over the screen cone (Fig. 13–1B), flatten and attach to the large hoop by means of rivets (Fig. 13–1, C and D). The screen cone is made of ½- or ¼-inch mesh hardware cloth. Cut a trial pattern out of newspaper to determine the exact shape of the wire. Figure 13–1E shows the half-circular layout of a cone where A and B give the measurement corresponding to A and B on Fig. 13–1B. Note that an extra ½ inch is added beyond the length (A to B) which is bent inside the hoop and is soldered in place, as shown in Fig. 13–1C. The measurement C to D on Fig. 13–1E represents the circumference of the large hoop. Leave enough extra wire beyond point D to overlap and strengthen the cone. Roll the cone, wire it together, and solder it in place in the large hoop.

Next, place the small hoop inside the large hoop and drill a ³⁄₁₆-inch hole through both hoops, as shown in Fig. 13–1B. Solder a ³⁄₁₆-inch nut inside the small hoop over the hole, as shown in Fig. 13–1C. The large and small hoops may be joined with short ³⁄₁₆-inch bolts. Almost any stout cloth, such as muslin, may be used for the net cone. Make your measurements of length (A to B) and diameter (B to D), as shown in Fig. 13–1E. Be sure to leave a surplus of cloth to go around the smaller hoop and form a hem (Fig. 13–1C). The net bag may be blind or may be open to receive a collecting vial, as desired.

This net is very durable, but should be washed promptly with fresh water and dried immediately after use. The towing wire should be fastened to the apex of the cone to give greater strength.

Other Common Methods.   The dip net, or water net (Chapter 12), is valuable for collecting single specimens from the water and for obtaining protective plants; the dredge net (Chapter 12) is also valuable. The pelagic forms of crustaceans, including larvae, are taken primarily with the coarse plankton net in open water. The night light (consult Chapter 1) used in marine waters is especially productive for small crustaceans.

Many species of small, fresh-water crustaceans produce thick-shelled resting eggs which withstand low winter temperatures or complete drying during droughts or summer periods. Therefore, the dry bottom mud from temporary ponds and pools may be rich with these so-called "resting" eggs (ephippia). To study the probable crustacean population (and that of other invertebrates) of such temporary ponds, scrape the upper ⅛ to ¼ inch of dry mud into a container, place this in an aquarium in the laboratory, and add sterile pond water (boil and cool) or distilled water. Within a few days to five weeks, many species of invertebrates will appear in such cultures. Small quantities of yeast, malted milk, or other suitable foods may be used if feeding is desired.

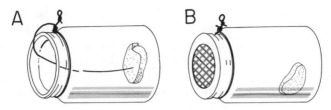

FIG. 13–2.   Two simple traps used with a meat bait for small crustaceans.

Narcotizing or Quieting Small Crustaceans

Small crustaceans may be examined directly under the microscope in depression slides. They may be slowed mechanically by adding methyl cellulose to the water, as described for the protozoans. Guyer (1929) suggests adding 2 parts of 1-percent Chloretone for every 5 parts of water in the culture. Specimens left briefly in this mixture may be revived by washing in clean, habitat water.

Killing and Preservation

Isolate specimens in a vial or small bottle and add 4 or 5 drops of 10-percent formalin to poison the animals. When all specimens have settled to the bottom, carefully remove the water and replace first with 35-percent and then with 50-percent alcohol, for 30 minutes each. Store in vials of 70-percent alcohol, with the collecting data included. Vials should be of screw-cap construction. Place vials in a larger, airtight bottle with the same preservative, for permanent storage.

The formalin-habitat water solution may be replaced with 3- to 4-percent formalin for both temporary and permanent storage, if desired. Some workers like to kill specimens by pipetting them directly into 70- to 80-percent alcohol for killing and preservation. This may cause some distortion of the fragile forms. Pennak (1953) recommends killing copepods directly in 95-percent alcohol and storing them in 70-percent alcohol.

Slide Mounts

Stains are not always essential for small crustaceans. Should stains be desired, acid fuchsin may be used at the water level (before specimens are placed in alcohol), or borax carmine may be used at the 70-percent alcohol level, as directed below. These stains are especially good when animals are treated in bulk. If appendages are to be dissected from larval and adult forms, especially from copepods, mount them in glycerin jelly, Monk's mounting medium in Permount, or similar media, as directed below. Pennak (1953) suggests that all copepod appendages may be seen if from five to ten whole specimens are mounted, ventral side up, on the same slide, for the appendages become positioned in such manner that all may be seen satisfactorily by examining the entire series.

Resinous Mounts Without Stain.   Although the author prefers to use stain, the present basic method will be presented, as it is satisfactory in itself and will serve as a basis for alternate mounting methods and staining techniques. Freshly killed specimens, or those preserved in formalin, should be slowly moved up through 35-, 50-, 70-, 85-, and 95-percent alcohol, and two washes of absolute alcohol, with 15 to 30 minutes per wash. Clearing may be carried out in clove oil, cedarwood oil, or xylene. Equal parts of absolute alcohol and the clearing agent should be used as an intermediate step into the clearing agent itself. At this point, check specimens under the microscope to determine if they are too transparent; if so, stain as directed below. Depression slides, slides with a spun cell or a plastic slide cell, or slides with supported coverslips may be required for thicker specimens (see Appendix B). Transfer a specimen to an appropriate slide and blot away the excess clearing agent, leaving only enough to prevent air bubbles from forming under the specimen. Add enough mounting medium (balsam, Permount, or similar mounting media) to fill the space under the coverslip, tease any bubbles away from the specimen with a fine needle, and add a coverslip by dropping it straight down on the mounting medium. Keep the slide flat until thoroughly dry. No ringing is necessary. Add labels with taxonomic and collecting data.

Staining Overcleared Specimens.   In a modified technique from French (1942), specimens which are too transparent may be stained while in the clearing medium by adding a few drops of light green, eosin, chlorazol black, or orange G stain to the clearing agent. To make these stains add excessive quantities of powdered stain to absolute alcohol, filter, and use. Overstaining is not likely to occur. This method may be used as a standard practice for all slide mounts, as well as an emergency treatment.

Rapid Mounting Method.   Specimens may be mounted directly from sea water, fresh water, or alcohol in Turtox CMC-10. This nonresinous medium can be used for killing, mounting, and clearing specimens quite satisfactorily. Small amounts of CMC-S may be added if stain is required (see Appendix B).

Alternate Mounting Options.   If resinous media are used, specimens may be transferred from 90-percent alcohol into Terpineol, which will dehydrate and clear them for mounting in balsam or Permount, thus eliminating the need for absolute alcohol. Alternately, specimens may be transferred from 90-percent alcohol into Euparal Essence until cleared. They are then mounted in Euparal (see Appendix B).

Acid Fuchsin Stain Technique.   Acid fuchsin (Appendix C) is an aqueous stain; thus, specimens are stained before they enter alcohol or are rehydrated (Appendix B) down to the water level and stained. This stain is soluble in water but becomes less soluble in increasing percentages of alcohol. It may be intensified in its action by acidifying the stain. Should overstaining occur, tap water or alkaline alcohol will destain the specimens to the desired density. Some experimentation may be required to determine the time required for staining. Place specimens in the stain from a few minutes to an hour or more, removing one to water occasionally to check its intensity. Remember that the stain becomes intensified and apparently darkened in the clearing process; thus, specimens should not be too dark. From this point, dehydrate through the alcohols and mount in a resinous or alternate medium, as described above.

Borax Carmine Stain Technique.   Dehydrate specimens to the level of 70-percent alcohol. Transfer to borax carmine (Appendix C) from 1 to 24 hours. Destain in acid alcohol until the specimens are light pink. Dehydrate up through absolute alcohol, clear, and mount in a resinous medium or by one of the alternate methods mentioned above.

Glycerin Jelly Mounts.   Glycerin jelly mounts have long been used for permanent and semipermanent slides of small crustaceans. The material need not be stained, but is rather transferred to a solution of 70-percent alcohol containing 5- to 10-percent glycerin. Place specimens in this fluid for five days to a week, until most of the alcohol and water has evaporated. During this period the glycerin impregnates the specimens slowly enough to prevent the collapsing of tissues. Next, select a suitable depression slide (or create a spun cell or mount a plastic ring cell, see Appendix B) so as to accommodate the thickness of the specimens. Place a piece of glycerin jelly large enough to fill the cell to a slight excess, and heat (do not boil) the slide gently over a flame until the jelly is liquefied. With a fine forceps or bristle transfer a specimen to the glycerin, work it down into this medium, and position it in the desired way. The yet-warm glycerin jelly should flow evenly over the specimen, creating a slightly convex surface. Put a ring of Murrayite around the cell and lower the coverslip directly, squeezing out the excess glycerin jelly with a soft cloth. If a flat slide, with or without coverslip supports, is used, cut the excess glycerin jelly from around the margin of the coverslip after it has hardened. With either type of slide, seal the coverslip all the way around with two or more coats of Murrayite or some other suitable cement. When the labeling is completed the slides may be stored, preferably flat, in a slide box.

CIRRIPEDIA:
THE BARNACLES

Narcotizing

Specimens may be narcotized in the following way: Place barnacles in finger bowls with sufficient sea water to permit complete expansion. When they are expanded and active, add a large pinch of Chloretone or menthol crystals and let stand undisturbed for 6 hours. Specimens are usually fully expanded and insensible to gentle probing at this time. Introduce a small drop of formalin with a pipette directly on the appendages of one individual. ·If there is no contraction, preserve by adding enough formaldehyde to make a 5-percent solution; then wash and transfer to alcohol, as described above. The author finds that epsom salts are usable but must be added in increasing amounts every 10 to 20 minutes, require longer than 6 hours as a rule, and often fail to properly anesthetize all specimens. Barnacles may also be overcome in collecting containers or sealed jars which have some organic material (such as the visceral mass of a sea cucumber) which will putrefy the water, use up the oxygen, and kill the barnacles in an expanded condition.

LARGER CRUSTACEANS:
LOBSTERS, CRAYFISH, CRABS, AND OTHERS

Collecting

General Methods.   The large crustaceans are more active at night than during the day. Fresh-water species are most easily collected by trapping, but may be taken with a dip net in shallow weedy areas in lakes, streams, and rivers. Marine forms are generally trapped, dredged, or collected by hand in the intertidal zone. These specimens hide under and around any kind of debris or rubble, live within burrows, hide within the sand, and in general occupy almost all possible microhabitats. The dip net (Chapter 12) is useful for securing free-moving specimens. Crabs and the like are easily secured by hand, though often they must be grasped by the back edge of the carapace or in such a manner as to pin both pinching hands against the carapace. On small and medium-sized crabs and crayfish you can position your fingers immediately behind the pinching arms, which will prevent the animal from reaching back far enough to pinch during collection. Always collect specimens into plastic buckets provided with large quantities of nonmucus-secreting seaweed to provide hiding places. Without such a baffle specimens will fight and tend to shed their appendages. Small pinnotherid crabs may be screened from gravel near the low-water mark in bays and estuaries, or secured from inside the mantle cavity of clams, mussels, and the like. Hermit crabs remain motionless when alarmed, but quickly give their identity away if the collector remains quiet and watchful. Pieces of bait (fish or meat) lowered into tide pools will lure shrimp from their hiding places in 5 to 10 minutes, at which time they may be netted. Larger crabs and lobsters are readily obtainable by skin diving and looking carefully in crevices along reefs, and similar places. Spider crabs and decorator crabs occupy kelp beds, pilings, weed-covered rocks, and the like. Branching coral and coral blocks should be cracked to obtain the wealth of species hiding there.

Traps.   Crustaceans are easily lured to traps containing fish, meat, liver, or other baits. Solid baits may be tied directly inside the trap, while small pieces of bait may be placed in a plastic bottle with holes drilled through the side (Fig. 13–3C). The building materials for traps must be considered. Fresh-water traps may be made of most convenient materials, whereas salt-water traps are quickly destroyed (in 1 to 2 weeks) when in constant use. Heavy-gauge, galvanized hardware cloth or screen will make excellent traps for occasional use, but is attacked by electrolysis very quickly. Untreated cotton netting will also be destroyed in a matter of days in the marine situation, but treated cotton lasts a long time. Nylon netting stretched over a meal frame proves very durable and quite satisfactory. Traps should be left in the salt water no longer than necessary, and last longer if thoroughly washed in fresh water and dried after using.

Most traps work on the funnel principle. One such trap is made of ½-inch hardware cloth 36 inches wide and 58 inches long (Fig. 13–3, A–E). Cut the hardware cloth as indicated (Fig. 13–3D) according to your dimensions, bend, and wire together. Cut a door opening in the top of the trap, and secure a door by wire rings. Other traps are shown in Fig. 13–4.

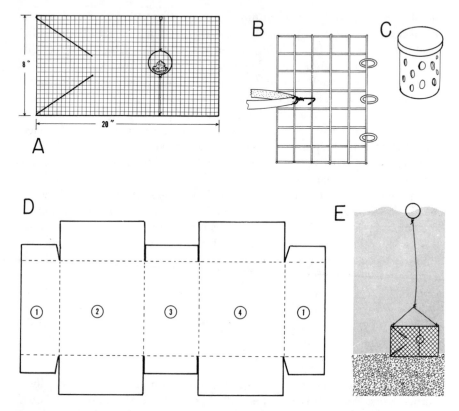

FIG. 13–3. A funnel trap used for small or large crustaceans. A. Side view of trap. B. Detail of door and rubber-band lock. C. Bait container. D. Pattern for cutting screen. (1) Funnel, (2) bottom, (3) end, (4) top. E. Using a float marker for locating the trap.

Preservation

Some Problems. Almost all of the large crustaceans (decapods) can autotomize or break off any of their five walking legs at will. Since shedding the legs is related to a defense reaction, these animals frequently lose one or all of their major appendages if placed directly in formalin or alcohol. In addition, muscle tissues will become brittle if improperly preserved. (The reader should consult the section on preservatives at the beginning of this chapter.)

Killing. All marine specimens may be killed by adding fresh water (50 to 75 percent) to the collecting containers, freezing in sea water (or

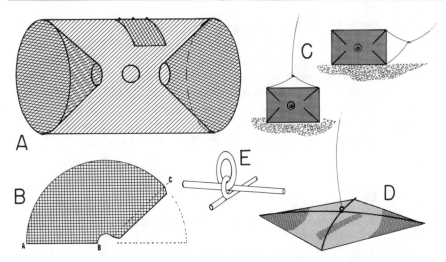

FIG. 13–4. A cylindrical funnel trap for crustaceans. A. Side view of trap. B. Cutting the funnel cone, where A to C equals the diameter of your screen cylinder. C. Setting the trap in quiet water or moving stream water. D. A manual crab trap. E. Detail of the cross-wires for the manual crab trap.

chilling tropical forms), being left out of water in sunlight, by isolating in small containers and adding 3 to 5 drops of formaldehyde which poisons specimens when left undisturbed, or by placing specimens in putrid sea water (containing rotting organic matter) which asphyxiates them. Either provide sufficient seaweed so that specimens may hide, or else separate them into small containers. Unseparated animals will fight and automatically shed some of their appendages.

Hermit Crabs.   Hermit crabs are prone to withdraw into their borrowed shells during preservation. They may be evicted by placing them in a mixture of fresh water and sea water. A few drops of formalin, insufficient to kill them, may hasten this process. If undisturbed by other individuals, hermit crabs will first withdraw into their shells under the stress of fresh water or formalin, but soon will leave their shells when withdrawal does not better their condition. Attempting to pull the crab free from its shell usually results in breaking off the abdomen. When the animal voluntarily leaves its shell, quickly transfer it to a mixture of salt and fresh water until it is dead. Some workers chip a hole into the mollusk shell and insert a curved wire, which generally causes the crab to move out of the shell.

Skeletal Studies.   For study of the complex maze of internal skeletal structures and calcified tendons, the soft parts of the tissue may quickly be reduced by injecting a 20-percent solution of potassium hydroxide (KOH) through the thin membranes of the body. As this reduces the internal tissues the carapace can easily be removed, at which time the entire animal may be submerged in a solution of KOH. As soon as most of the tissues are removed, wash the animal thoroughly in several changes of fresh water. The legs present a definite problem here and may be punctured at the membranes and submerged for several hours in each wash. Finally, rinse, dry, and store the specimen in a small box with its label.

REFERENCES

Darwin, C., 1851, *A Monograph on the Sub-class Cirripedia,* Roy Soc., London, Vol. 1.

Darwin, C., 1853, *A Monograph on the Sub-class Cirripedia,* Roy Soc., London, Vol. 3.

Edney, E. B., 1954, Woodlice and the Land Habitat, *Biol. Rev.* 29:185–219.

French, A. J., 1942, Notes on the Preservation, Mounting and Staining of Entomostraca, *The Microscope* 5:20–22.

Galtsoff, P., *et al.,* 1959, *Culture Methods for Invertebrate Animals,* Dover, New York.

Guyer, M. F., 1929, *Animal Micrology,* Univ. of Chicago Press, Chicago.

Kiser, R. W., 1950, *A Revision of North American Species of the Cladorceran Genus* Daphnia, Edward, Seattle, Wash.

Knudsen, Jens W., 1958, Life Cycle Studies of the Brachyura of Western North America, I. General Culture Methods and the Life Cycle of *Lophopanopeus leucomanus leucomanus* (Lockington), *Bull. South. Cal. Acad. Sci.* 57(1):51–59.

Knudsen, J. W., 1966, *Biological Techniques: Collecting, Preserving, and Illustrating Plants and Animals,* Harper & Row, New York.

Monk, C. R., 1938, An Aqueous Medium for Mounting Small Objects, *Science* 88: 174.

Morrison, R., 1943, Aqueous Media for Microscope Slides, *Turtox News* 21(1): 23–24.

Pennak, Robert W., 1953, *Fresh-Water Invertebrates of the United States,* Ronald, New York.

Richardson, H., 1905, A Monograph on the Isopods of North America, Washington, *U.S. Nat. Mus. Bull.* 54:i–liii + 727.

Van Name, W. G., 1936, The American Land and Fresh-Water Isopod Crustacea, New York, *Bull. Amer. Mus. Nat. Hist.* 71:iii–vii + 535.

(See Chapter 1 for other references.)

CHAPTER 14

Other Arthropods: Spiders, Scorpions, and Their Allies

ARANEAE: THE TRUE SPIDERS

Habitat and Collecting

Spiders occupy almost all terrestrial habitats, and a few have invaded the aquatic habitat. They are mostly nocturnal in behavior. Not all spiders build large nets between bushes; some hunt on the ground or hide in ambush on flowers, others live in the deep grasses and make irregular nets on the surface, and still others make cone-like webs extending back into burrows in the ground. Thus, the many species are extremely varied in structure, habitat, and especially in their behavioral patterns of feeding and self-protection. The number of people studying spiders is small compared to those interested in insects. This may be due to the fact that insects can be displayed more easily and in a much more attractive manner. Nevertheless, the study of spiders and other arachnids can hold a lifetime of interest.

Some Collecting Materials. The collector should carry a number of small jars and one or two large, wide-mouth (preferably plastic) jars with

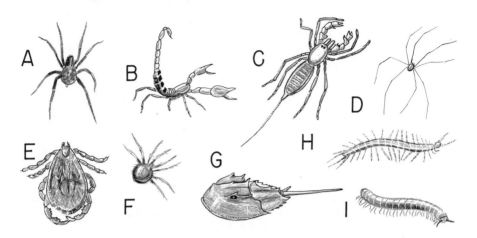

FIG. 14–1. Representative types of other arthropods. A. Spider. B. Scorpion. C. Whip scorpion. D. Harvestman, or long-leg spider. E. Tick. F. Water mite. G. Horseshoe crab. H. Centipede. I. Millipede.

70-percent alcohol. A small aquarium net is useful for snatching spiders from their webs before they have a chance to escape. Specimens may be picked up directly from the ground and placed into alcohol, caused to drop into alcohol from their webs and hiding places or quickly netted and then transferred to the alcohol solution. Field notes should be taken at the time of each collection, including the locality data (Chapter 1) and notes describing the nature of the web, the hiding place, and other critical factors concerning natural history. Killed specimens may be transferred in the field to smaller vials or bottles, with a field number which corresponds with the notebook description. The author uses very small paper bags for collecting live specimens. Bottles will do just as well, but they are quite heavy. Two spiders should not be kept alive in the same container. A specimen is netted and allowed to drop into the bag. The bag should then be closed, the top rolled down two or three turns and stapled shut. When large numbers of living spiders are to be collected, drop the bags into a cardboard box attached to a pack board. This will prevent crushing which might occur in a knapsack.

The Tuning-Fork Technique. A tuning fork or similar vibrating object is extremely useful, in addition to a small aquarium net, for capturing all kinds of web-building spiders. If the tuning fork is twanged and then placed against the web, spiders will be lured to the point of vibration, just as to a vibrating fly. Many spiders are extremely quick to detect that the tuning fork is not a fly and will race right back to their hiding place.

Nevertheless, if the aquarium net is held in readiness before the tuning fork touches the web the spider may be captured before it again reaches the safety of a crack or burrow.

Sweep Net.   The sweep net is a most useful tool for working through tall grass and bushes where spiders hide. The construction and use of this net are described in Chapter 12. Basically, this heavy net is beat into bushes from which hiding spiders fall into the net bag and are captured.

Sorting Pans.   Immature and small species of spiders often hide in leaf litter beneath trees, or in similar debris. A shallow, plastic sorting pan, at least 8 by 12 inches, is excellent for obtaining these animals. When quantities of such debris are sorted, small spiders can be seen running on the white surface of the pan. These may be picked up by forceps or, more readily, by a small water-color paintbrush freshly dipped in alcohol.

Preservation

Spiders are usually killed and preserved in 70- or 80-percent alcohol. If large or numerous specimens are placed in one jar, the alcohol should be changed after 24 hours. Five-percent formalin may be used in place of alcohol, though it is not preferred. Specimens intended for critical internal dissection should be preserved in FAA (Appendix C). Color does not preserve well and ultimately will be altered. Color notes and records may be made, however, as described in Chapter 1 (p. 15). Occasionally, spiders are placed in Riker mounts (see Chapter 12) where they are kept in a dry state. Owing to the extreme amount of shrinkage that occurs in the abdomen, however, it is best to draw off some of the abdominal fluid with a hypodermic needle and replace it with latex rubber (which is used to inject the veins and arteries of vertebrates). Alternately, the abdomen may be opened from the ventral surface and the visceral mass removed and replaced with cotton. Specimens are then placed in the Riker mounts and permitted to dry normally.

Collecting Webs

The collection of spider webs has long fascinated specialists. There are some rather awkward and unsatisfactory techniques for mounting two-dimensional spider webs between sheets of glass or on paper treated by ink. The author has devised a method which is extremely simple, quick, and very satisfactory. The materials needed are an aerosol spray-paint can containing a dull black, oil-base paint or lacquer, a 1-inch paintbrush with soft bristles, a jar of three parts of Elmer's Glue and two parts water (thoroughly mixed), and some heavy white-surfaced Bristol board or posterboard cut to appropriate sizes. Thinner cardboard and heavy paper are less satisfactory, as will become obvious. Locate a well-formed, two-dimensional spider web, preferably before it is destroyed by insects. Chase the spider from the web by blowing on it, but note carefully where it hides. Now spray the net from a distance of about 18 inches, using circular strokes. The

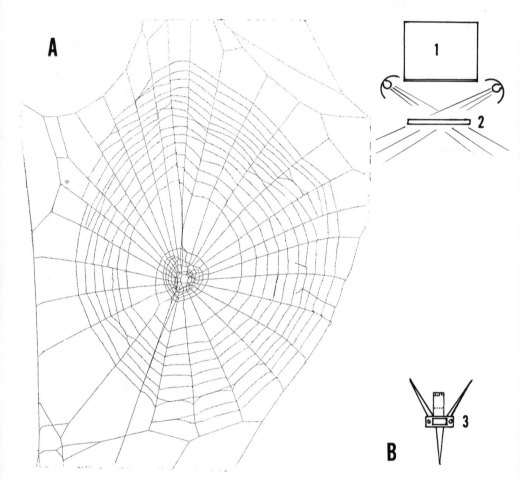

FIG. 14–2.   Spider-web techniques (see text). A. The center of an actual spider web which was collected on a sheet of cardboard and is here reproduced at its original size. B. Looking down on a setup for photographing spider webs in the laboratory. (1) A blackened box or velvet cloth, (2) the spider web supported on a frame or piece of vegetation, (3) the camera supported on a tripod.

spray must be a fine mist rather than a splatter and should be directed in such a manner that all of the web is evenly blackened, but none of it becomes loaded with heavy drops of paint. The result of such a spraying is that very fine particles of wet paint cling to the spider web. Now, select a suitable piece of Bristol board large enough to contain the central portion of the web. Paint the entire surface of the board with a moist coat of glue (avoid an excess of glue). The surface should be tacky, but not dripping wet, as the spider web will slip on the surface. Next, carefully maneuver the board behind the web and then bring it forward in such a manner that the entire web is pressed into the wet glue. With your fingers, or with a pocket-knife, carefully break each of the supporting guide "wires" where they leave the cardboard. Be careful not to move the cardboard up and down, or from side to side, after the cutting has begun, as the web will be pulled out of its true symmetry. Place the collected net in the shade where it may dry, and then record the field number or other data at the bottom. Figure 14–2 is made of an actual spider web which was collected in this manner. Such webs are permanent and may be stored in filing boxes. If they are to be subjected to heavy usage, a coat or two of artist's spray fixative, used as directed on the can, will render them very durable.

SCORPIONS, PSEUDOSCORPIONS, SUN SPIDERS, AND OTHERS

Habitat and Collecting

Scorpions.   These animals (Fig. 14–1B) are found in the semitropical southeastern United States and the arid southwestern United States extending north through eastern Washington into British Columbia, and south into Mexico and elsewhere. They are generally thought of as desert animals and occupy various habitats, from brushy foothills to sparsely vegetated sand dunes. Like all nocturnal animals, scorpions must locate some suitable hiding place for daytime dormancy. Many of them occupy self-made burrows; others crawl under any debris such as bark, rocks, fallen vegetation, and the like. Until you are familiar with scorpions, gloves or forceps should be used for picking these animals up. There are a few dangerously poisonous species in the Southwest. These are quite small over all, with pinching hands which are smaller in diameter than the wrists.

Among the sagebrush hills in the Southwest scorpions make burrows which are somewhat oval at the mouth (Fig. 14–3A). The burrow extends

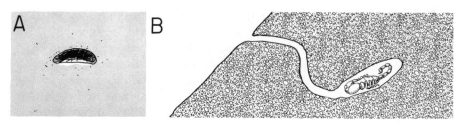

FIG. 14–3.   Scorpion burrow. A. Mouth of the burrow. B. Cross-section of the burrow.

back into the hillside and then curves, as diagrammed in Fig. 14–3B, the scorpion hiding in the chamber at the end. When digging burrows, push a piece of soft grass down through the tunnel so that it may be located should it fill with dirt. Begin digging by cutting a hole in front of and below the burrow level. Then with a pocketknife chip away the dirt of the burrow so that it falls into the depression below. This way the burrow will remain clean and open and there will be no danger of encountering the scorpion before you are ready to. As you chip back through the burrow the scorpion will eventually drop down into the depression and may be captured.

Sun Spiders.   Sun spiders, whip scorpions, and the like, occupy much of the same range as the true scorpions. Many of these, like scorpions, are located hiding under temporary shelters or may be picked up under street lights or by walking the sand dunes with a lantern. These animals may be handled by forceps or with a gloved hand, although, like scorpions, they may be picked up with the bare hands once the collector is thoroughly familiar with their behavior. Since it is very easy to become lost on desert sand dunes at night, have one member in your party (never work alone) drag a heavy stick to make a uniform mark which will serve as a trail back to your automobile.

Pseudoscorpions.   Pseudoscorpions (Fig. 14–1C) are common the world over, from the tropics on up into the higher latitudes. They are small and secretive and frequent leaf litter, grassy fields, sea shores, nest cups in rodent burrows, or any place where food (notably soft-bodied insects and mites) is available. In some parts of the country they are even house guests. Leaf litter and other debris may be sorted in pans to isolate the specimens or may be treated with the Berlese funnel (see Chapter 12, p. 136).

Preservation

These animals should be preserved in 70-percent alcohol (see the discussion under "Preservatives" in Chapter 13). They may be killed by dropping directly into this or some lesser solution of alcohol. In the latter case they should be transferred to 70-percent alcohol. If numerous animals or large individuals are preserved, change the alcohol solution after 24 hours. Pseudoscorpions may be mounted on slides intact or dissected.

TICKS

Habitat and Collecting

Juvenile and adult ticks will either be found on host animals (mammals, reptiles, and occasionally birds) or waiting on vegetation to attach to some host animal. (Hosts must be secured by methods given in Chapters 18–20.) Rodent runways and deer trails are sometimes alarmingly alive with adult ticks. The sweep net (Chapter 12) is excellent for securing these specimens, or the tick drag net (Baker and Wharton, 1952) will also give good results. The drag is made of flannel cloth 1 yard square tacked on one side to a stick 36 inches long. A short piece of rope is tied to the stick, on either end, to form a loop which the collector pulls along bushes and grass near animal trails. The drag should be inspected periodically for ticks which cling to it.

Preservation

Ticks may be collected into, and preserved in, 70-percent alcohol. They should be kept in vials along with their field data which, in turn, are enclosed in large, airtight jars filled with preservative.

Slide Mounts

It is not always essential to make slide mounts of ticks for taxonomic work. Also, ticks need not be treated in potassium hydroxide (KOH) unless such treatment is desired. This treatment removes all of the soft tissues, permits dorsal and ventral details to be observed simultaneously, and allows the specimen to be flattened somewhat more easily. Otherwise, ticks should be killed in a flat position between two slides.

To macerate, treat in a 10-percent KOH solution overnight and check periodically until the soft tissues are reduced. The alternative method is to use a stronger solution in the same manner, or to gently heat the specimens for a matter of minutes in 10-percent KOH. When the tissue is removed, wash specimens in two changes of 35-percent alcohol, the second being acid alcohol. Dehydrate up through absolute alcohol, clear in xylene, and mount in balsam or Permount. Specimens may be mounted directly from water or alcohol, whether macerated or not, in Turtox CMC-10. Stain is generally not needed for ticks. Larval ticks should be treated like mites (see below).

MITES
AND WATER MITES

Habitat and Collecting

Mites are small and fragile animals which require special care. They are best collected and preserved in small vials of 70-percent alcohol. The simplest way to capture and transfer specimens to alcohol is by dipping a small, pointed, water-color brush in alcohol, touching the specimen, which will then adhere to the brush, and washing it off in the alcohol.

Parasites may be collected from beetles and all land vertebrates (see Chapters 19 and 20) or from nesting materials occupied by these animals. Place fresh-killed hosts in white enamel pans and comb the hair or feathers in search of mites. Mites will be seen walking across the pan and can be captured with a paintbrush. Nesting material should be processed in the Berlese funnel (see Chapter 12, p. 136). The alternate method is to place such material in large white enamel pans and carefully pick through it. If the debris is in small quantities, mites may be seen walking on the debris or on the pans. White paper plates may be placed on grass and observed periodically for small, pinkish-colored mites which are present. Finally, a Birge's cone net (Chapter 13) is excellent for obtaining water mites from ponds and lakes where they occur in heavy vegetation. A small aquarium dip net is useful for this same purpose.

Preservation

Mites are best preserved in 70-percent alcohol in small vials containing the field data. Screw-cap or corked vials are preferable to cotton-stoppered vials because of the small size of the animals. Vials should be placed in an airtight jar filled with preservative. Slide mounts are useful and are often made for critical taxonomic study.

Slide Mounts

Gray (1964) recommends two types of mounting media for mites: "(1) A high refractive-index medium like Berlese's for the very heavy-walled forms, such as the Oribatid mites and the pseudoscorpions and (2) a low refractive-index mountant like Gray and Wess', for the thin-walled forms, such as the Tyroglyphid and Gamasid mites." Place a suitable amount of the proper medium (Appendix C) on a blank slide and transfer a live or preserved mite with as little water or alcohol as possible. Press the specimen down in the mounting medium, expel air bubbles, and add a coverslip. Label and dry in a flat position. When thoroughly dry, ring the slide with a waterproof cement such as Murrayite, ringing varnish, or other.

Baker and Wharton (1952) in their excellent text on ticks and mites observe that "workers at the University of California have developed a methocellulose formula which has proved to be excellent for many mites. They found the best procedure was to clear thoroughly in lactophenol before mounting, although some of the more delicate mites needed no special preparation as the lactic acid in the medium cleared the specimens sufficiently." See Appendix C for the methocellulose medium.

In addition to these media mites may be mounted in Permount or balsam by treating the heavy-bodied forms with KOH as described for the ticks, moving all forms up through the alcohols, using 35-, 50-, 70-, 85-, and 95-percent, and absolute alcohol, two changes at 10 to 15 minutes each. Transfer to a mixture of half absolute alcohol and half clove oil and, finally, to pure clove oil for clearing. When the specimen is sufficiently clear transfer to a clean slide, add the desired resinous medium, lower a coverslip into position, and dry. Add labels with both the locality and host data in the taxonomic data.

MEROSTOMATA

These strange-appearing animals (Fig. 14–1G), known as horseshoe crabs or eastern king crabs, are found intertidally and in pools above sandy beaches on the eastern seaboard. Specimens are readily captured by hand and picked up directly by the long, sword-like telson (tail). Although they are not related to crustaceans, the body is constructed of similar materials, permitting these organisms to be killed and preserved exactly as the large crustaceans (see Chapter 13).

CENTIPEDES
AND MILLIPEDES

Habitat and Collecting

Centipedes and millipedes (Fig. 14–1, H and I) are broadly distributed; they are common in the tropics, desert and arid regions, grasslands, and forests. Centipedes are almost exclusively nocturnal, whereas millipedes may be active either during the day or at night. They retreat into hiding places such as under stones, wood, downed vegetation, leaves, or the burrows of other animals. Some establish home sites, whereas others find new hiding places at random. In higher latitudes these animals are commonly found overwintering in rotten logs.

Collecting should be conducted by searching likely hiding places. The gloved hand or a pair of forceps suffices to capture these animals. Many species are mildly poisonous; few are really dangerous. Nevertheless, treat all as if they were poisonous, for safety's sake.

These animals may be killed in an insect-killing bottle (Chapter 12) or by dropping into alcohol. Place fresh-killed animals in pans, arrange the legs and straighten the body, and add 70-percent alcohol. After a few days transfer specimens to vials or museum jars with their field data; keep them in 70-percent alcohol. Specimens intended for dissection should be preserved directly in FAA.

For study collections, specimens may also be pinned in the manner of insects, but with less satisfactory results. They may be arranged and glued to cardboard which is in turn pinned, or they may be pinned directly with a temporary cardboard support, as directed in Chapter 12.

REFERENCES

Baker, Edward W., and G. W. Wharton, 1952, *An Introduction to Acarology,* Macmillan, New York.

Baker, W. W., *et al.,* 1958, "Guide to the Families of Mites," Contribution No. 3, Institute of Acarology, Univ. of Maryland, College Park.

Comstock, J. H., 1948, *The Spider Book,* Comstock, Ithaca, N.Y.

Crompton, J., 1954, *The Life of the Spider,* Mentor (New American Library), New York.

Gertsch, W. J., 1949, *American Spiders,* Van Nostrand Reinhold, New York.

Gray, Peter, 1964, *Handbook of Basic Microtechnique,* McGraw-Hill, New York.

Kaston, B. J., and E. Kaston, 1971, *How to Know the Spiders,* Brown, Dubuque, Iowa.

Knudsen, Jens W., 1956, Pseudoscorpions, A Natural Control of Siphonaptera in Neotoma Nests, *Bull. South. Calif. Acad. Sci.* 55(1):10.

Knudsen, J. W., 1966, *Biological Techniques: Collecting, Preserving, and Illustrating Plants and Animals,* Harper & Row, New York.

Petrunkevitch, A., 1933, The Natural Classification of Spiders Based on a Study of their Internal Anatomy, *Conn. Acad. Sci. Trans.* 31:299–389.

Savory, T. H., 1964, *The Arachnida,* Academic, New York.

(See Chapter 1 for other references.)

CHAPTER 15

The Echinoderms:
Sea Stars, Sea Cucumbers,
and Others

The five classes of echinoderms commonly overlap in their distribution; thus, collecting techniques will be presented for the phylum as a whole. A number of problems arise in preservation. (1) Because of the calcium carbonate skeleton, echinoderms cannot be left in any acid preservative. On the other hand, (2) many echinoderms can be dried and stored more economically than by preservation in liquid. However, some of the echinoderms, notably the starfishes, are identified on the basis of muscle attachments for which dried material is unsatisfactory. Many of the echinoderms have (3) the tendency to break off "arms" when subjected to harsh preservatives; (4) others, notably the sea cucumbers, are highly contractile, and at least some representative specimens should be preserved in a relaxed state, with the tentacles expanded.

The more essential echinoderm techniques are presented herein. However, the parent textbook, *Biological Techniques* (Knudsen, 1966), provides techniques for working with larvae, tube feet, pedicellaria, histology, and storage cabinet construction.

COLLECTING ECHINODERMS

Brittle stars, serpent stars, starfishes, sea urchins, sand dollars, sea cucumbers, and sea lilies occupy intertidal and deep-water habitats with both rocky and sandy substrates. The feather stars (crinoids) are exceptions; they spend part of their time as swimming animals. Hand-collecting and dredging (or dragging) are the two chief methods of collection.

Drags and Dredges

In deeper water with sand, mud, or rock-rubble bottoms, the starfish drag or dredge proves excellent for collecting. The drag (Fig. 15–1A) utilizes string mops which entangle starfishes and brittle stars. String mops may be attached to any suitable object, but the pattern described here is recommended. To make a starfish drag, make a 90° bend in the center of a solid 6-foot bar of metal or heavy pipe. Weld a ring at the bend to receive a tow rope, and four additional rings as shown. Attach a 4-foot piece of heavy, galvanized chain, by means of a split ring, to each of the four rings. Finally, attach two or three string mops to each chain, as shown. To operate, simply attach a rope to the tow ring, pay out sufficient line for the depth of the water, so as to allow the drag to work flatly along the bottom, and tow

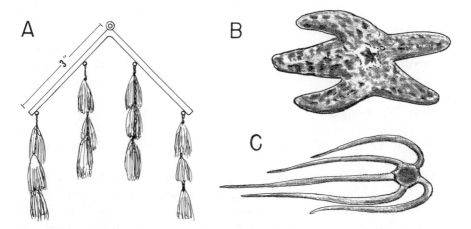

FIG. 15–1. Echinoderm techniques. A. A starfish tangle containing string mops. B. The desired position of large starfish specimens intended for drying. C. The position for drying brittle stars.

from a skiff. In suitable habitats collectors will be amazed at the volume of organisms obtained.

The biological dredge (see Chapter 1, p. 4, for construction and use) will retrieve a fantastic number of echinoderms. Sort dredge hauls or drag hauls immediately, and place specimens in buckets of clean sea water. Avoid overcrowding, and exchange the sea water frequently if the specimens must remain some time before preservation.

Hand-Collecting

Skin diving, wading in the water, and collecting during periods of low tides are all satisfactory for echinoderms. On sandy bottoms the tentacles of burrowing sea cucumbers denote where to dig; starfishes and brittle stars are exposed on the bottom; sand dollars bury themselves, but leave telltale rings in the sand where they are buried. In the Puget Sound area, tremendous concentrations of sand dollars are found on sandy beaches near where some source of fresh water enters the sea water.

When collecting on rock and sand bottoms, or on pure rock substrates, look for the tentacles of sea cucumbers projecting into the water from between rocks or from the sand just underneath protective rocks. Others that may be exposed by turning rocks are small starfishes, sea urchins, and brittle stars. Many of the large starfishes seek cover from the receding tides in order to avoid drying. Thus, they are usually found in the more shaded and protected crevices. Tidepools along rocky shores provide many echinoderms, especially sea cucumbers and brittle stars.

Specimens should be placed in buckets of moist, nonmucus-secreting seaweed. Periodically, fill the bucket with fresh sea water and drain again. Sea cucumbers should not be crowded or mixed with other animals, in that they tend to eviscerate or slough off the outer skin, and thus quickly decompose. Sea cucumbers should be preserved as quickly as possible.

One should not overlook pilings and floating docks, for starfishes and sea cucumbers are especially numerous there, hiding among other organisms such as mussels and barnacles.

Always bear in mind the method of preservation and the size of containers available. There is no value in collecting extremely large specimens for general use, especially if the collector has no container which will accommodate large specimens. Although size is less a factor for dried specimens, it should, nevertheless, be considered at the time of collecting to avoid wasted specimens.

PRESERVATION TECHNIQUES

Classes Asteroidea and Ophiuroidea: The Sea Stars and Others

Museum specimens are generally dried or preserved in liquid; specimens for classroom dissection must be preserved in liquid. Rapid-killing techniques may be tried, but narcotizing is recommended for starfishes, and is essential for brittle stars and serpent stars since these animals will break off their appendages.

Narcotization and Liquid Preservation. Both epsom salts and fresh water may be used to narcotize and kill these animals, but epsom salts are greatly preferred. Here, crude epsom salts seem to work much better than the chemically pure form of magnesium sulfate. Place specimens in flat-bottomed pans with just enough sea water to cover them. Add a small handful of epsom salts (depending somewhat on the volume of water) every hour during the day, and let stand overnight. After this, test the specimens, first by gently probing them, and second by introducing a drop of formalin near the specimens. This will cause contraction or squirming if the specimens are not ready for preservation. When they are ready, transfer the specimens to a fresh pan of water and arrange the arms so that they take up the least amount of room (Fig. 15–1B). Brittle stars and serpent stars are always positioned as follows: Place the specimen in a pan of water, put your finger on the disc and move the animal in a direction away from the longest arm, so as to position it as shown in Fig. 15–1C. The longest arm should be perfectly straight for the measurements and plate counts essential to taxonomy. When the specimens are arranged, add enough formalin (preferably neutralized) to make a 5- or 10-percent solution, and let stand for 24 hours.

From this point specimens are either dried or preserved in alcohol. For liquid preservation, wash the specimens to remove the formalin, transfer to 50-percent alcohol for 1 hour, and store in 70-percent alcohol. Isopropyl alcohol (70 percent) may be used in place of ethyl alcohol for echinoderms.

The fresh-water technique is essentially the same as that described above, except that specimens are placed in pans of fresh water from 6 to 12 hours or more. After this they are arranged and preserved as above.

Drying Methods. Echinoderms make excellent dried specimens when fixed in formalin, as described above. Untreated specimens may be dried directly, but with less success for larger specimens. Dried specimens tend to retain their color quite well if stored in a dark container. Air movement, rather than heat, is the key factor in drying. Thus, moderate heat with sufficient air circulation is preferable to hot ovens or direct sunlight. Arrange specimens on screen wire to permit air circulation above and below. The author finds the attic of the science hall an excellent, warm place for drying. Specimens are placed on screen near the discharge of the air ducts. One to three weeks may be required for complete drying of large specimens. However, this process can be hastened with a drying oven, described below.

Drying Oven.   A very convenient dryer can be cheaply built which will meet all of the requirements for drying echinoderms and many other types of biological specimens (Fig. 15–2). The heating element may consist of a 200-watt bulb, a heating coil, or a hot plate, as desired. The dryer may be of any dimension, so long as the basic principle is adhered to.

The materials needed are: two side pieces of $\frac{3}{4}$-inch plywood, 18 by 42 inches; one top piece of $\frac{3}{4}$-inch plywood, 18 by 24 inches; one back piece of $\frac{1}{4}$-inch plywood, 24 by $40\frac{3}{4}$ inches; two (door) pieces of $\frac{1}{4}$-inch plywood, 12 by $40\frac{3}{4}$ inches; fourteen drawer runners, 1 inch by 1 inch by 18 inches (note, these will measure $\frac{3}{4}$ inch by $\frac{3}{4}$ inch); fourteen tray ends, 1 inch by 1 inch by $17\frac{3}{8}$ inches; fourteen tray sides, 1 inch by 1 inch by $22\frac{3}{8}$ inches; one lower front brace, 1 inch by 2 inches by 24 inches; seven pieces of $\frac{1}{4}$-inch hardware cloth, $17\frac{3}{4}$ by $22\frac{1}{4}$ inches; two pairs of butt hinges and one small hook and eye.

First, attach the drawer runners to the inside of the side pieces, 5 inches on center, measuring from the top, as shown in Fig. 15–2A, using glue and nails, as shown in Fig. 15–2D. Now, nail the top piece to the two side boards, as shown in Fig. 15–2C. Drill four $\frac{3}{4}$-inch holes about equally spaced in the upper part of the back plywood piece (Fig. 15–2A, fourth hole hidden on left), and two holes in the top of each door piece (Fig. 15–2B). Glue and nail the back piece to the sides and top, making sure that the cabinet is square. Mark and cut the lower front edges of the side boards to accommodate the front brace (Fig. 15–2E). Glue and nail the front brace in position. Attach the two doors so that they overlap the top piece (Fig. 15–2B). Finally, install the hook and eye, or other door catch, as desired. Cut the ends of the drawer pieces so that they will dovetail, as shown in Fig. 15–2E. Glue and nail the trays, square the assembled pieces, and install the hardware cloth on the bottoms (preferably with cleats, or simply with large-headed tacks).

If a light bulb is used as a heat source, obtain a porcelain light receptacle, a 4- by 4-inch electrical outlet box, an appropriate length of lamp cord,

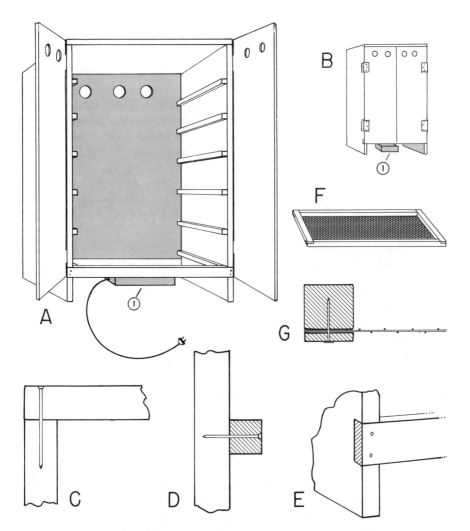

FIG. 15–2. A simple drying box useful for echinoderms or other biological specimens such as mushrooms. A. A front view of the drying box. B. Front view with the doors closed. (1) Heating element. C. Detail of top and side assembly. D. Adding the shelf cleats. E. Detail on installing the lower front brace. F. The finished screen shelf. G. Detail of attaching the screen to its frame by means of a cleat.

and a male wall plug. It may be necessary to shield the light bulb from dripping preservatives by placing a sheet of tin directly above it on the bottom screen. If a light is the heat source, the seventh or bottom screen will probably be omitted.

To use this dryer simply narcotize and fix specimens as directed above, permit them to drip off excess preservatives for a few moments, and then place them on the trays.

Skeletal Techniques.   The soft epidermal tissue may be removed from starfishes to display the skeletal structure. Select small specimens (3 to 4 inches in diameter) or portions of larger specimens. Macerate the tissues by placing the entire specimen in straight Clorox, Purex, or a 10-percent solution of sodium hydroxide. Remove the specimen from time to time, wash, and then scrub vigorously with a toothbrush to determine if the tissue is ready for removal. It is preferable not to let maceration go to a point where all tissues are removed, as the skeletal plates may become disassociated. When the tissue will readily slough off, clean externally and flush the internal contents from the body by removing the oral membrane and the stomach. Thoroughly wash in two changes of fresh water, for ½ hour each, and in one change of acid 30-percent alcohol (Appendix C). Dry and store in a small cardboard box with a lid.

Storage of Specimens.   Over long periods of time, echinoderm colors will remain more stable if the specimens are stored in the dark. Dried echinoderms will attract dermestid beetles and other insect pests if they are improperly stored. A storage cabinet with a rubber or plastic seal to prevent the invasion of insects is recommended.

Class Echinoidea: Sea and Heart Urchins, Sand Dollars

Liquid Preservation.   Although urchins can be suitably dried, some should be preserved in liquid, as the structure of the mount parts, the so-called "lantern," and the associated muscle attachments are of taxonomic importance. Specimens may be killed and preserved in 70-percent alcohol or killed in 5-percent formalin, washed, and preserved in 70-percent alcohol. The alcohol solution should be changed after 24 hours. Isopropyl alcohol is suitable in place of ethyl alcohol. As soon as larger specimens are placed in the preservative, arrange the spines in such manner that they will more easily fit into a jar. Inject specimens intended for dissection through the oral membrane, to ensure rapid internal preservation.

Positioning Spines for Preservation.   Frequently, urchins are desired for wet or dry specimens with the spines positioned in a lifelike attitude. Usual methods of preservation often cause the drooping of spines. You can obtain lifelike specimens by placing the urchin in a container (a coffee can will do) with ample sea water. When the animal has fully erected the spines in the desired position, quickly sift fine sand all around the animal until it is completely buried (Fig. 15–3, A and B). With your hand or some flat object hold the sand in place and pour off the excess water. Replace this with 10-percent formalin, let stand for 24 hours until the muscles holding the spines harden, and then gently wash away the sand. The specimens should then be washed thoroughly in fresh water and preserved in alcohol, as described above, or dried directly without any washing. If drooping persists in large specimens during the drying process, simply pour off all the fixative and dry the specimen in the container, sand and all. After drying is completed, the sand can easily be removed.

Drying Methods. Dried specimens are easily maintained, require no expensive preservatives, and retain their color quite well. However, the spines will become broken if the specimen is mistreated. It has long been the practice to remove the visceral mass by cutting around the oral membrane (Fig. 15–3C), removing the lantern and scooping out the viscera with a wire. This procedure is unnecessary, however, as the viscera contain a high percentage of water and dry very well if they have been previously preserved. Likewise, it is desirable to keep the lantern intact for future reference. The procedure, therefore, should be preservation followed by drying, as described for the starfishes above.

Class Crinoidea: Sea Lilies and Feather Stars

These fragile echinoderms live below the low-tide line, on down to the deep-sea habitat. They are therefore seldom sought, but are taken inci-

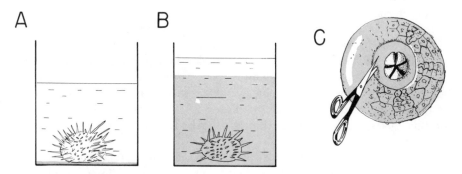

FIG. 15–3. Techniques for working with sea urchins. See the text for a detailed explanation.

dentally during standard dredging. The inclusion of one or two sea lilies is always the high point of a dredge haul.

Liquid Preservation.   Crinoids are prone to break up during preservation. Bianco (1899) recommends killing specimens by plunging them directly into 90-percent alcohol, after which they should be stored in 70-percent alcohol. The author has not preserved crinoids, but would assume that narcotizing in epsom salts, as described for the starfishes, would be desirable prior to preservation. Some specialists relax crinoids in the dark (they contend that light causes crinoids to contract) and hold the specimens flat to prevent movement until they are killed by 95-percent alcohol. Quick freezing with CO_2 or slow freezing in the refrigerator may also be attempted.

Drying Methods.   Place killed crinoids in a mixture of 2 parts glycerin and 3 parts 50-percent alcohol for 24 hours, remove, dry, and store in a small cardboard box, as described above.

Class Holothuroidea: Sea Cucumbers

Sea cucumbers require special treatment from the time they are collected until the time preservation is completed. They tend to slough off the outer epidermis and eviscerate as soon as they are vaguely disturbed. Therefore, preserve specimens as quickly as possible.

Narcotizing and Preserving.   One method for relaxing cucumbers is to place them in sea water until they are expanded, and then to add small quantities of epsom salts every 30 minutes over a period of many hours, increasing the amount each time. It may be essential to let them stand overnight until they are completely insensible to touch. At this point, kill in 10-percent formalin, wash, and preserve in 70-percent alcohol, or kill them directly in 70-percent alcohol. Inject the body cavity with preservative. Replace the alcohol after 24 hours.

An alternate method that may be tried is to permit the animal to expand completely in sea water. At this point grasp the animal with a pair of wooden forceps where the tentacles join the body, lift the body with the opposite hand, and submerge the tentacles in concentrated acetic acid. Simultaneously, an assistant should inject large quantities of 95-percent alcohol through the anal opening by means of a syringe. The anal opening is then plugged and the animal is placed in 70-percent alcohol. Variations of this method are injection through the body wall or through the oral aperture.

When adequate time is not available for the careful relaxation of specimens, simply preserve directly in 10-percent formalin or 70-percent alcohol.

Such specimens are the least desirable, but their use is preferable to wasting the animal. Sea cucumbers must never be left in formalin unless intended solely for classroom observation. Formalin destroys the dermal ossicles which are found in the skin and are used taxonomically.

Dermal Ossicles.   Remove a piece of cucumber skin and treat it in 10-percent KOH or full-strength Clorox or Purex. This reduces the dermal tissue, leaving the ossicles. These should be thoroughly washed in fresh water and stored in a screw-cap vial of alcohol, along with the field data and specimen identification. Such vials are, in turn, sealed in an airtight jar filled with 70-percent alcohol.

REFERENCES

Bianco, S. L., 1899, The Methods Employed at the Naples Zoological Station for the Preservation of Marine Animals, *Bull. U.S. Nat. Mus.* 39(M):1–37.

Clark, A. H., 1921, Sea-lilies and Feather Stars, *Smithsonian Misc. Coll.* 72(7): 1–43.

Fisher, W. K., 1911–1930, Asteroidea of the North Pacific and Adjacent Waters, *Bull. U.S. Nat. Mus.* 76(1–3):1–1020.

Galigher, A. E., 1934, "The Essentials of Practical Microtechnique," published privately.

Gray, Peter, 1964, *Handbook of Basic Microtechnique*, McGraw-Hill, New York.

Hyman, L. H., 1955, *The Invertebrates: Vol. 4, Echinodermata*, McGraw-Hill, New York.

Knudsen, J. W., 1966, *Biological Techniques: Collecting, Preserving, and Illustrating Plants and Animals*, Harper & Row, New York.

Mortensen, T., 1927, *Handbook of Echinoderms of the British Isles*, Oxford Univ. Press, New York.

(See Chapter 1 for other references.)

CHAPTER 16

The Fishes

For simplicity, the scope of this chapter has been broadened to include the cartilaginous fishes such as the sharks and rays; the bony fishes, which include the familiar game and food fish; and the cyclostomes, referred to as lampreys and hagfish. Although some of these groups are more or less widely separated taxonomically, techniques of collection and preservation are generally similar and permit common treatment in this text.

The parent textbook, *Biological Techniques* (Knudsen, 1966), provides additional fish techniques and construction methods for otter trawls, electrical fish shockers, Fyke nets, and other fish traps.

One should be aware of the taxonomic features of any group of organisms, and specimens must be preserved in a way that yields maximum value.

Body measurements are also critical in taxonomic and growth studies. Figure 16–1 shows nine of the more common measurements: some of these are the total length from the snout to the end of the tail, the standard length from the snout to the end of the vertebral column, the predorsal length from the snout to the first dorsal fin, the head length from the snout to the back of the operculum.

Finally, adequate field data must be taken if museum specimens are to be of all-around value. This must include not only the geographic locality

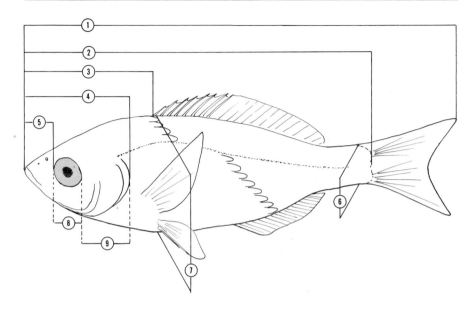

FIG. 16–1. Standard fish measurements. (1) Total length, (2) standard length, (3) predorsal length, (4) head length, (5) snout length, (6) least height of caudal peduncle, (7) greatest body height, (8) ocular length, (9) post-orbital length.

of the collection and the date (see Chapter 1), but should include a description of the habitat in which the specimens were collected, color notes taken from the living specimens (see Chapter 1), and the like. Specimens that are well preserved lend themselves to internal dissection and studies of the food content of the stomach.

COLLECTING METHODS

Laws and Regulations

Write to your department of fish and game to determine the state laws concerning fish collecting and other types of collecting. Almost all states require fishing licenses and many require additional collecting permits. These permits can easily be obtained and often require little more than simply notifying the department of fish and game. Any fish specimens may be taken during the proper season, if they are covered by the state fishing license. However, although persons bearing a special permit are entitled to more freedom in collecting than the general angler, few states permit the use of rod and reel, nets, or traps in closed waters or during closed seasons unless the collector is accompanied by a game protector. Likewise, the use of rotenone or other chemical poisons for fish collecting is prohibited in most states unless specific authorization is obtained from the game protector. Although these laws may be slightly cumbersome, the serious student or teacher is not in any way inhibited from conducting his scientific research. Such laws make for a greater degree of plant and animal conservation and, therefore, should be honored.

Construction and Use of Fish Nets

Dip Nets.   The construction of a lightweight dip net is described in Chapter 12. For working with fishes a larger mesh in the netting is required as well as a more sturdy hoop. Round iron rod measuring ¼ or ⅜ inch in diameter proves quite suitable. The rod should be bent and fastened to the handle as explained in Chapter 12. For a dip-net bag, a so-called "smelt and minnow" replacement dip-net bag (or other replacement bag available commercially) is most satisfactory. On the other hand, ⅛- or ¼-inch square minnow-seine netting may be fashioned into a dip-net bag in the manner described in Chapter 12. One problem that develops is the abrasion of the bag where it goes over the hoop. Fish dip nets are therefore usually threaded with wire, as shown in Fig. 16–2A, and the net bag is then applied to this. Stefanich (1952) suggests drilling ⅝-inch steel airplane tubing, which is used for the hoop, and threading steel wire (size 14) in such a way that it occupies the inner diameter of the hoop (see Fig. 16–2B). When the dip-net bag is threaded to the wire, this type of construction allows the bag to be reversed for rapid emptying or cleaning.

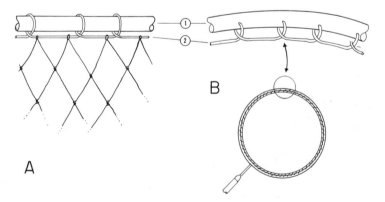

FIG. 16–2.   Methods of making dip nets. See text for complete explanation. (1) Net hoop, (2) wire.

Seines.   The seine net is perhaps one of the oldest and most useful types of net for general collecting, especially for smaller fishes. Minnow seines usually are 4 feet deep, or more, and may measure from 6 to 20 feet or more in length. They are equipped with a lead-sinker line on the bottom and a float line on the top, and usually have a pair of sticks attached to either end to keep the seine open. The netting may be of cotton construction, although nylon is much preferred as it is practically rot-resistant. The two most popular mesh sizes are ⅛- or ¼-inch square.

A simple minnow seine is made as follows: After determining the size of the net required, select a suitable mesh. Square meshes are hung on a 1:1 basis, or 1 foot of netting applied directly to 1 foot of top or bottom line. On the other hand, diamond-hung stretch mesh is hung on a 3:2 basis where 18 inches of stretched netting are hung per foot of top or bottom line. The top and bottom lines may be made of ³⁄₁₆- or ¼-inch cotton or, preferably, nylon line. Tapered hardwood floats, 3 by 1½ inches, are spaced at 2-foot intervals along the top line. Tubular lead seine weights, 1½ inches long by ½ inch outside diameter, are distributed at 1-foot intervals along the bottom. Stout nylon or cotton twine should be used to apply the net, floats, and sinkers to the top and bottom lines. Cut the top line several feet longer than required for the net. Attach one end of the line to a tree at chest height, and then thread the required number of floats on to the line. Tie the opposite end of the line to some object also at chest height. Firmly attach a piece of cord, preferably waxed with bees wax, to the top line and begin attaching the net and floats as shown in Fig. 16–3, A, B, and E. Square mesh may be applied with a running-chain stitch rather than the clove-hitch (Fig. 16–3E). Repeat the procedure when hanging the bottom line with its leads and netting. When larger seine nets are being made, the seine weights may be replaced by a heavy galvanized chain applied firmly to the bottom line, as shown in Fig. 16–3D. This will give added strength to the seine and make it fish on the bottom along its entire length. Seines measuring longer than 20 feet are best hung in this fashion. Finally, a pair of sticks 4 feet or longer should be attached to either end, to facilitate handling the net.

Figure 16–3F shows a typical lake or pond set. One worker pulls his end of the net far out into the water while the second collector remains close to shore. The end of the net is brought around in a wide arc and, finally, both ends are brought up on the beach. As the net moves into shallow water it tends to bag out to trap the small fishes.

A typical beach set along sandy-bottomed ocean or lake shore may be made with a large seine, as shown in Fig. 16–3, G and H. Two lines 300 or 400 feet in length are attached to bridles on either end of a 100-foot seine. This seine should be equipped with a chain along the bottom, as mentioned

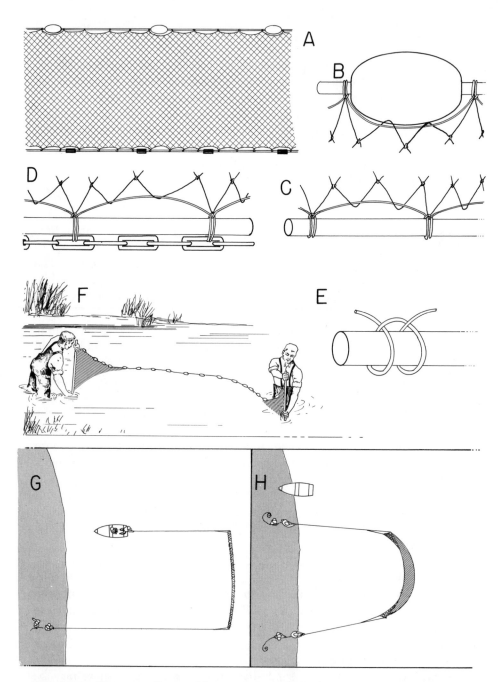

FIG. 16–3. Hanging and using fish nets. A–E. Making the minnow seine (see text for complete instructions). F. Using the minnow seine. G. Setting a large beach seine offshore by means of a skiff. H. Retrieving the beach seine in such a way that it remains parallel with the shore (see text for details).

above. The net is loaded into the skiff as follows: One side rope is payed into the skiff hand over hand, followed by the net, and in turn, by the second side rope. A collector holds the end of the first side rope on the shore while two men in a skiff row directly away from the shore, paying out rope until they reach the seine. They then move parallel to the beach until all of the seine has been payed out, and then they row in toward shore, paying out the second side line. The seine should be given adequate time to sink to the bottom before it is retrieved toward the shore. By putting a marker on the ropes every 50 feet the two teams pulling in the seine can notify each other of their progress and thus keep the seine parallel to the shore (Fig. 16–3H). The seine is brought in at only moderate speed, in order to keep fishes moving ahead of it, rather than greatly alarming the fishes and causing them to swim around the ends of the net. When the net reaches the shore the fishes are removed into buckets and other containers.

Surface Trawl.   Chapoton (1964) developed the following surface trawl for catching juvenile American shad. The trawl net is made of ½-inch stretched mesh nylon netting; it measures 4 by 12 feet at the opening, is 15 feet long, and tapes to form the cod end as shown in Fig. 16–4A. The two ends of the trawl are attached to a frame made of 2- by 2-inch pieces of wood which are strengthened at the corners with ½-inch plywood. The head rope of the trawl has the usual floats, whereas the foot rope is supplied with lead weights. A short bridle is attached to either end of the trawl frame.

Chapoton used 5/16-inch polyethylene rope for pulling the surface trawl. The main tow rope is 25 feet long and has a pulley, which attaches to the bridle of the boat, at one end and a snap hook at the other. Two additional pieces of polyethylene rope 50 feet long are equipped with snap hooks at either end. One is attached to either bridle on the trawl frame and both are snapped into a ring at the end of the tow rope, as shown in Fig. 16–4.

A 16-foot aluminum boat with a 10- or 18-horsepower outboard motor is used to pull the trawl. To operate, the trawl is floated away from the boat. When it is trawled, the net uprights into a "fishing" position. At the end of the trawl, the wooden frame tilts backward and the lead-line immediately buoys to the surface with the frame, thus preventing the escape of caught fish. (From Chapoton, 1964, p. 144.)

Care of Nets.   Any net or line should be washed in fresh water immediately after it is used. Fish and other marine animals and plants contaminate nets with mucus which will begin to rot and will, in turn, destroy the fabric. Salt water alone is sufficient to damage cotton netting if it is not quickly washed and dried. Nylon netting tends to withstand much more abuse, can be kept wet for long periods of time, and requires less cleaning

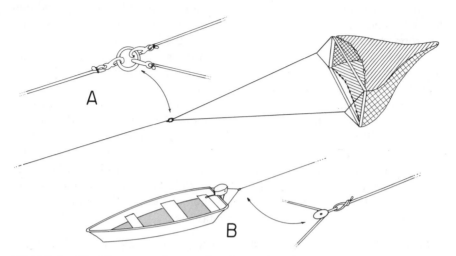

FIG. 16–4. The Chapoton surface trawl, redrawn after Chapoton (1964). A. A diagram of the surface trawl and a detail of the lead lines. B. A detail of attaching the trawl to the skiff. See text for complete instructions.

than cotton netting. Cotton netting should be periodically treated with tar, copper sulfate, or other preservatives. Consult Graumont and Wentstrom (1948) or other good texts for procedures of treating and making nets.

Fish Traps

Minnow Trap. Figure 13–4 (p. 187) illustrates a funnel trap which is appropriate for both crustaceans and fishes. The basic principle of the minnow traps is the funnel which leads toward a bait. Fishes enter the trap in an attempt to get at the bait and then find it difficult to escape. The trap may be square or round or any form one chooses. One funnel is often sufficient, but two or more may increase the catch. Figure 16–5 illustrates a collapsible minnow trap made of ½-inch stretched nylon mesh. This trap utilizes two rings of No. 9 galvanized wire, 1½ inches in diameter, and two rings of No. 9 galvanized wire, 10 inches in diameter and provided with four small eyes (Fig. 16–5, D and C, respectively). Four green willow sticks serve to hold the trap open (Fig. 16–5, A, D, and E). Each funnel is hung between its 10-inch ring and its 1½-inch rings. The funnels are then assembled and the small rings are tied together, as shown in Fig. 16–5D. Next, a piece of netting is applied between the 10-inch hoops to form the outside wall of the trap. A cut in this mesh forms a door which is closed by a piece of twine (Fig. 16–5A). If the four supporting sticks are removed, this trap will collapse and store flat. When the trap is put into operation the sticks are placed in position, the trap is baited and is lowered into a stream or pond by means of an anchor string. The bait generally consists of chopped fish tied in cheesecloth and suspended between the two funnels.

Fish Poisons

The use of fish poison is controlled by law, as mentioned above. The most common compounds contain rotenone and come in either liquid or powdered form. These are diluted with water and are used primarily in tide pools and similar areas where the poison will eventually become diluted. Such materials should not be used in small ponds, simply because a total kill of the pond will eventually result. Rotenone and similar compounds eventually kill gill-breathing invertebrates as well as fishes. For this reason the collector must always bear in mind the need for conservation while collecting. Rote-

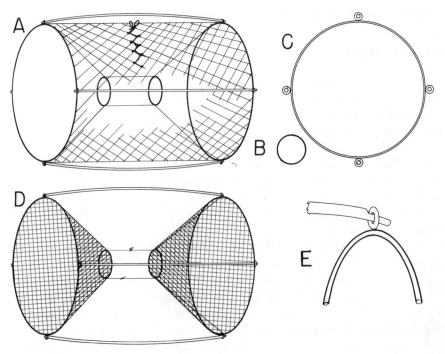

FIG. 16–5. A collapsible minnow trap made of fish netting. A. A side view of the trap. B–C. The supporting rings. D. Tying the funnels in position. E. Detail of inserting the spacing sticks.

none works better in warmer water temperatures; it becomes less effective and requires a longer period of time in cold water. A new liquid product known as "Chem Fish Collector" has been developed (Chemical Insecticide Corporation, New Jersey) which is not affected by water chemistry or temperature; the manufacturers claim that it works more quickly than simple rotenone.

Manufacturers of fish poison recommend a base dilution for their products; the poison is diluted and added to the pool to be collected. Within a few minutes fishes will begin to float to the surface and can be retrieved by means of a dip net.

Quinaldine Live-Catch Method

It is the habit of many reef dwellers to enter rock or coral crevices or burrows when approached. A skin diver may simply squirt quinaldine (see Appendix C) into their hiding places from a plastic squirt bottle. This anesthetizes the fish and causes it to "stagger" out into dip net range. Fish quickly revive in the dip net and then are transferred to plastic bags.

Other Methods

Skin divers should consider spear fishing as an excellent means for obtaining specimens selectively. The spear gun is less satisfactory than a spear propelled by rubber tubing. The latter, especially when equipped with a barbless, multipronged wire spearhead, does very little damage to average-sized specimens. Speared fish should be placed in plastic bags immediately, due to shark danger. If sharks are present, send fishes to the surface by releasing air into the bag, tying, and allowing it to ascend. A spear net is illustrated in the parent text, *Biological Techniques* (Knudsen, 1966).

Slurp guns are widely used for small reef fishes, especially those that live in burrows. While these are commercially available, a simple slurp gun may be built from a 42-inch length of clear rigid plastic tubing with an inside diameter of 1¼ inches plus or minus. A loosely fitting rubber stopper equipped with a wire serves as a piston. With the piston near the mouth of the slurp gun, the collector simply puts the gun near to a small fish, pulls back on the plunger and thus sucks his specimen into the clear plastic tubing. Commercial slurp guns usually have a much larger diameter cylinder and piston arrangement to which a smaller barrel is affixed. This permits a greater volume of water to be sucked into the gun and ensures fish capture.

TRANSPORTATION
OF LIVE FISHES

Some simple techniques for the transportation of live organisms and the control of temperature, metabolic waste, oxygen concentration, and the like, are given in Chapter 1. Small fishes may be transported in polyethylene bags filled with water (if the bags are kept cool), as the bags tend to permit the exchange of gases. However, the science of live-fish transport has developed far beyond the scope of methods needed by a student or teacher. Nevertheless, the reader should be aware of some of these developments. There is a large literature dealing with tank design, insulation, and handling techniques; more concerning circulation and pumping devices, aeration by spray and compressed-gas liberators, etc.; still other literature dealing with filtration equipment, temperature control, and the like. Anesthetics are now widely used in transporting fishes. Some of these are quinaldine, chlorobutanol, ether, methanesulfonate, urethane, sodium amytal, and thiopental sodium. Three of the best recent papers which deal with these techniques give tables concerning the effect of anesthetics and present lengthy bibliographies of additional literature: McFarland (1960), Norris *et al.* (1960), and McFarland and Norris (1958).

PRESERVATION TECHNIQUES

Killing and Hardening

Kill fish specimens rather than letting them die in the air. Unless otoliths are required, add a small quantity of formaldehyde to a large container of water and place the fishes in this. If the container is large enough the specimens will die perfectly straight. Smaller fishes should then be transferred to bottles with 5-percent formalin. Fishes longer than 5 or 6 inches should have a cut made in the right side of the abdominal cavity and should be put in 10-percent formalin for hardening. Those fishes that will not readily fit into jars (extremely long individuals, for example) may have to be bent and tied with string during the hardening process. One must always bear in mind the size of containers ultimately available. Extremely large specimens should be rejected when more convenient smaller ones are available. Fishes may also be killed by putting them into 70-percent alcohol while they are still alive. Regardless of the preservative, if the fishes are dead the slime should be removed from the body prior to preservation. Many workers feel that fishes required for skeletons should never be placed into formalin, as softening of the bones may occur. Consult Appendix B for notes on the use of MS-222 fish anesthetic.

Shipment of Field Specimens

When a large number of field specimens has been amassed and preserved in formalin, wrap individual specimens or lots of specimens in cheesecloth, along with their field data. Pack these into metal containers or tin cans (see Chapter 1) in such a way that no movement of the specimens will be possible. Wet the specimens with the initial preservative, but pour off the excess, and seal the container. If the container is sealed properly the specimens will not become dried and will travel very nicely for periods of many weeks. The critical thing is to be sure field data are included with each lot of specimens.

Soaking Out and Final Preservation

If field specimens preserved in formalin are to be transferred to alcohol for final preservation, most workers first remove the formalin before transferring to alcohol. This is done simply so that the fish specimens will not reek of formalin when handled later on. Place specimens in bottles of clean water and change the water each day. Two or three days (or more) may be required to remove most of the formalin. Some workers watch for a telltale scum to form on the surface of the water as a sign that the soaking-out process is completed. This indicates that the formalin is now so weak that bacteria can begin to grow. At this time specimens should be placed in 70-percent ethyl alcohol or 40- or 50-percent isopropyl alcohol. If a large volume of specimens is placed in one container the alcohol should be changed after 24 hours.

Specimens are kept in tightly sealed museum jars, along with their field data. The preservatives always modify the color; light will further bleach the specimens. Thus, it is advisable to store specimens in a dark place.

Color Preservation

Probably no solution yet devised will preserve fish color (or that of other animals) perfectly. This is one notable area in which research is still required. Many methods have been attempted, including preservation in gasoline. Martin and Terrence (1962) experimented with ten different solutions for color preservation in fishes. At the time of their publication the specimens had been preserved only 7 months, which is significant but not quite conclusive. The solution that showed the most promise for them is made as follows: 20 cubic centimeters of 1-percent phenol, 12 cubic centimeters of formaldehyde, 8 cubic centimeters of glycerin, and 60 cubic centimeters of water. At the time of their writing, both the color and the animal tissue were in a good state of preservation.

Kotthaus (personal communication, 1963) comments that small numbers of fishes preserved in an arsenic-glycerin solution in 1950 were not yet rendered colorless at the time of writing. His preservative is made as follows: 30 cubic centimeters of glycerin, 6 grams of potassium acetate, 0.08 grams of sodium acetate, 0.012 grams of arsenic trioxide, and 63 cubic centimeters of water. This fluid becomes dark with time, but tends to maintain color longer than direct formalin-preserved specimens.

Otoliths

Otoliths, fish earbones, are becoming increasingly important in fish studies, for they show growth rings and are morphologically different on the generic level and, thus, are useful for aging specimens and for taxonomy. Otoliths should be removed from fresh-caught fish or from those killed and preserved in alcohol. Fishes that have been in formalin, even for a few hours, are not satisfactory in that the fine characteristics on the otoliths are quickly eroded.

To remove the otoliths cut through the skin on the side of the head and remove the opercular bone. Next, shave away the thin bone making up the capsule of the ear. When exposed, the otolith can be removed with a forceps. Occasionally, the otoliths will fall down into the brain cavity. Should this occur, simply go on with the removal of the otolith from the opposite side and then flush the "lost" otolith from the brain cavity by means of a stream of water directed through the ears. Otoliths have very little tissue clinging to them and may be placed directly into small dry vials, along with their data. Screw caps rather than cotton stoppers should be used, since the fine sculpturing on the otolith may become entangled in the cotton.

Gut Contents

If the gut contents are to be examined for food studies one should recognize the following: (1) Immediate preservation of the contents must be achieved by injecting preservatives into the gut or by removing the gut for special preservation. (2) Formalin may destroy the calcareous shells of food organisms and thus should be avoided if these shells are important.

SKELETAL TECHNIQUES

The study of fish skeletons is one of the keys to fish taxonomy. Smaller fishes, up to 10 or 12 inches, should be cleared and stained unless they are rather deep and thick-bodied. Elongated specimens up to this length are very suitable for this technique. Bigger fishes are usually prepared by der-mestid beetles, which do a very satisfactory job. Specimens may also be boiled and hand-picked, but are less satisfactory. Because these techniques are almost identical for all vertebrates they will be discussed collectively in Chapter 21.

REFERENCES

Arnold, E. L., 1951, An Impression Method for Preparing Fish Scales for Age and Growth Analysis, *Prog. Fish-Cult.* 13(1):11–16.

Butler, R. L., and L. L. Smith, 1953, A Method for Cellulose Acetate Impressions of Fish Scales, With a Measurement of Its Reliability, *Prog. Fish-Cult.* 15(4): 175–178.

Chapoton, R. B., 1964, Surface Trawl for Catching Juvenile American Shad, *Prog. Fish-Cult.* 26(3):143–144.

Clemens, W. A., and G. V. Wilby, 1946, Fishes of the Pacific Coast of Canada, *Bull. Fish Research Board of Canada,* No. 68.

Crowe, W. R., 1950, Construction and Use of Small Trap Nets, *Prog. Fish-Cult.* 12(4):185–192.

Eddy, S., 1970, *How to Know the Freshwater Fishes,* Brown, Dubuque, Iowa.

Graumont, R., and E. Wentstrom, 1948, *Fisherman's Knots and Nets,* I–XV, Cornell Maritime Press, Cambridge, Md., 1–203.

Hubbs, Carl, and Karl Lagler, 1949, *Fishes of the Great Lakes Region,* Cranbrook Institute of Science, Bull. No. 26, Bloomfield Hills, Mich.

Knudsen, J. W., 1966, *Biological Techniques: Collecting, Preserving, and Illustrating Plants and Animals,* Harper & Row, New York.

Lewis, W. M., 1963, *Maintaining Fishes for Experimental and Instructional Purposes,* Southern Illinois Univ. Press, Carbondale, Ill.

Martin, M., and M. Terrence, 1962, A Solution for the Preservation of Color in Fishes, *Turtox News* 40(4):122–123.

McFarland, W. N., 1960, The Use of Anesthetics for the Handling and the Transport of Fishes, *Calif. Fish and Game* 46(4):407–431.

McFarland, W. N., and K. S. Norris, 1958, The Control of pH by Buffers in Fish Transport, *Calif. Fish and Game* 44(4):291–310.

Norman, J. R., 1963, *A History of Fishes,* Hill & Wang, New York.

Norris, K. S., *et al.*, 1960, A Survey of Fish Transportation Methods and Equipment, *Calif. Fish and Game* 46(1):5–33.

Rayner, H. J., 1949, Direct Current as Aid to the Fishery Worker, *Prog. Fish-Cult.* 11(3):169–170.

Rounsefell, G. A., and W. H. Everhart, 1953, *Fishery Science: Its Methods and Applications,* Wiley, New York.

Rupp, R. S., and S. E. DeRoche, 1960, Use of an Otter Trawl to Sample Deep-Water Fishes in Maine Lakes, *Prog. Fish-Cult.* 22(3):134–137.

Schrenkheisen, Ray, 1938, *Field Book of Freshwater Fishes of North America North of Mexico,* Putnam, New York.

Schultz, L. P., 1936, Keys to the Fishes of Washington, Oregon, and Closely Adjoining Regions, *Univ. Washington Publ. Biol.* 2(4).

Schultz, L. P., and E. M. Stern, 1948, *Ways of Fishes,* Van Nostrand Reinhold, New York.

Scott, W. B., and W. H. Carrick, 1967, *Freshwater Fishes of Eastern Canada,* Univ. of Toronto Press, Toronto.

Smith, S. H., 1954, Method of Producing Plastic Impressions of Fish Scales Without Heat, *Prog. Fish-Cult.* 16(2):75–78.

Stefanich, F. A., 1952, An Improved Dip Net, *Prog. Fish-Cult.* 14(4):172.

Walford, L. A., 1937, *Marine Game Fishes of the Pacific Coast from Alaska to Ecuador,* Univ. of Calif. Press, Berkeley, Calif.

The Amphibians

CHARACTERISTICS

Before collecting and preserving amphibians the collector should have a thorough knowledge of those characteristics of taxonomic importance, so that his specimens will be preserved to yield the maximum amount of information. In addition to gross morphology, ornamentation, and the like, body measurements, skeletal structures, color in life, and even voice are important. Figure 17–1 shows some of the common measurements made on salamanders and frogs. *Note,* the snout-to-vent length is the standard length.

It is imperative that specimens be preserved in a uniform way, with the body straight, the legs uniformly arranged, and feet flat in salamanders, in order to facilitate measurements and study. Amphibians which have been dropped directly into strong formalin are so grotesquely twisted and out of "posture" that they are both unattractive and difficult to work with. Thus, if specimens are to be preserved at all, the collector should be prepared to relax and fix his specimens properly, to take color notes while the animals are still alive, to record the voice if this is essential, and so on.

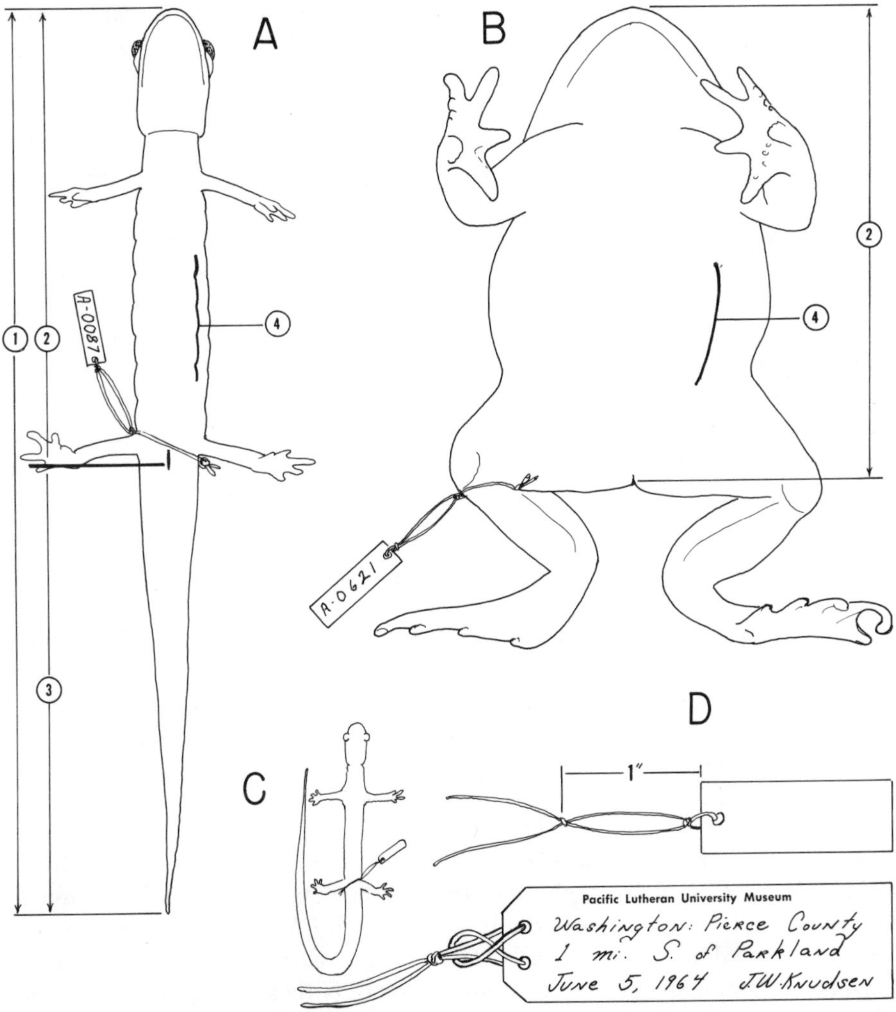
A
B
C
D
A-0087
A-0621
1"
Pacific Lutheran University Museum
Washington: Pierce County
1 mi. S. of Parkland
June 5, 1964 J.W. Knudsen

FIG. 17–1. Measuring, tagging, and posturing amphibians. A–B. Measurement and tagging procedure for salamanders and frogs, respectively. C. Posture of long-tailed salamanders. D. Method of tying the tag and an enlargement of a typical leg tag. (1) Total length, (2) snout-to-vent length, (3) tail length, (4) cut here or inject with formalin. See text.

FIELD DATA

Make a practice of recording complete field data in a notebook at the time of collection. This should include not only the date and geographic locality (see Chapter 1), but should also have such things as the time of day, the weather conditions, air and water temperatures, the general habitat, the microhabitat in which the collection was made, and notes concerning the activity or lack of activity of the specimens. All of these data help to create a picture of amphibian natural history. A field notation such as "collected in woods" is of little value as compared with "collected in woods, north-facing slope, heavy overgrowth with 90-percent coverage, specimens under rocks and heavy leaf litter only in areas with ground-water seepage." Many collectors record either ground or water temperature and rectal temperatures of the specimens. Any unusual observations, such as breeding activity, egg-laying, feeding, and so on, should be recorded. If possible, color notes should also be recorded, either in the field or immediately after returning to the laboratory. (See Chapter 1 for techniques of color notations.) Field notes are usually arranged by collecting station, each station having its own individual number. This same number will accompany both the specimens from that station and the logging of specimens in the museum.

NEED FOR CONSERVATION

There is a notable tendency on the part of collectors of amphibians and reptiles to overcollect in areas of low population density. For example, spade-foot toads which live in desert and semidesert situations will migrate en masse to temporary waterholes after the first heavy rain during the breeding season. Owing to the harshness of such habitats, the population is naturally quite low. Thus, overzealous collectors may destroy the total population by simply taking ten or fifteen specimens. This is especially true where the microhabitat of amphibians is discontinuous, confined to small areas which have some unique physical factors that make amphibian existence possible. Such situations are usually well enough isolated to make reoccupation of the habitat quite unlikely. Conversely, where the physical requirements of certain species are continuous, such as those of the common tree frog *Hyla regilla* which is continuous from central California up into British Columbia on the west side of the Cascade Mountain range and elsewhere, there is little chance of overcollecting by taking moderate numbers, since reoccupation and migration are always possible. Usually there are few laws pertaining to the collection of amphibians. However, some states, such as

California, require a collector's permit for every kind of plant and animal. These are readily obtained from fish and game departments and usually cost about 1 dollar per year.

COLLECTING METHODS

Equipment and Field Problems

The kind of equipment used to capture amphibians will depend on the habitat. Collecting containers, which are required in large numbers, present one of the basic field problems. Specimens must be isolated first by their collecting station, and second by species or, at least, genera. Almost all amphibians secrete mucus which may be poisonous to other species. Specimens cannot be overcrowded in containers and must be kept both moist and cool. Therefore, the field car should be parked in shade and specimens should be kept in the car itself rather than in the trunk, which will quickly overheat in warm weather. In collecting from a car, cartons of quart and pint museum jars make excellent containers. Collect specimens into plastic bags or plastic bottles provided with wet paper towels, moss or leaf litter. About ½ inch of water should be placed in each jar, along with a small quantity of paper toweling for frogs. For salamanders, loosely crumple up one or two paper towels and put them in a bottle. Thoroughly saturate the towels with water before adding the specimens. On longer field trips the paper towels should be changed daily and the bottles washed out in order to remove mucus and excreta. Place a complete set of field notes and a field station number, written on a good grade of paper with a medium-soft pencil, in each bottle, along with the specimens. Specimens will require no food for long periods of time if kept cool. Do not screw the lids down so tightly that oxygen cannot enter the bottle. Keep the bottles in their pasteboard cartons or styrofoam boxes to provide darkness for the specimens.

Frogs and Salamanders

Ponds and Lakes.   Some species of frogs and salamanders are permanent residents in ponds and lakes and thus can be found during all favorable seasons. The presence or absence of large predaceous frogs such as the bullfrog will frequently determine the presence or absence of other species. Almost every pond and lake has its breeding population that migrates from the terrestrial habitat during its particular breeding season. Migrations may start as early as the time when the winter snows leave the ground (about January 15 in the Pacific Northwest) and will continue up into the early summer months, each species taking its turn. Salamanders, tree frogs, and toads can be collected during this migration or found hiding under boards, logs, and so on, near the breeding pond. Use a coarse-meshed dip net with a moderately long handle for netting frogs and salamanders. Fine-meshed dip nets are difficult to move through the water rapidly and are less suitable for amphibians. The minnow seine (see Chapter 16, p. 213 for construction and use) is excellent for collecting adult and larval amphibians. A small triangular dredge, such as that described in Fig. 1–1E, is excellent for sampling ponds and small lakes. Figure 17–2, A, B, and C, shows how a small dredge on 100 feet of line can be carried around the end of a small pond, released, and retrieved in order to catch larval and adult forms.

Collecting at night with a gasoline pressure lantern or flashlight is quite satisfactory. Frogs that are "singing" can be cautiously approached and netted. When approaching a singing male frog use the light only during periods when it is singing. When it stops singing, switch the light off and wait quietly until singing resumes. Frog eyes will glow in the beam of a flashlight directed parallel with the surface of the water. It is often necessary to swat the dip net directly down over the frog and pull the net in toward shore in one continuous motion. Techniques that may work for frogs during daylight hours are the use of a trout fly to literally hook the frog or the use of a .22-caliber pistol with bird shot for large bullfrogs.

Minnow traps (see Chapters 13 and 16) baited with fresh-cut liver will often attract aquatic salamanders. During the breeding season an unmated female salamander isolated in a small screen cage within the minnow trap will usually lure dozens of males (and also females) with the chemical secretions given off in the water. The author has had unbaited minnow

FIG. 17–2. Steps in sampling a small pond with a hand dredge.

traps that caught salamanders for days after an unmated female had been present in the trap. Frog, toad, and salamander eggs may be found singly or in chains or clumps on the bottom of shallow ponds or attached to vegetation. When collecting eggs, keep them isolated in plastic bags of water or jars of water. Attempt to select egg masses with different rates of development, in order to preserve a complete developmental series.

Rivers and Streams.   Tropical salamanders and frogs are usually terrestrial or arboreal, whereas those that live in the temperate zones are more commonly found in rivers and streams. Large aquatic salamanders, such as the northwestern *Dicamptodon,* are easily caught on hook and line or in baited minnow traps. When collecting along streams and rivers, slowly turn debris such as logs, loose stones, and the like, in search of adult animals. Night collecting will often reveal considerable amphibian activity where both adults and larval forms move to the shallow waters for feeding. Specimens that live in swift streams usually swim rapidly downstream when alarmed. Therefore, they should be approached from a downstream position to ensure their capture. Many tropical frogs live on the vegetation just above small streams and rivers. They can be heard calling and can be located by flashlight at night. Some species attach their eggs to vegetation hanging directly over such streams. There the larval forms develop and, upon hatching, fall into the stream to complete their growth. Where the terrain permits, a seine can be quite useful for catching amphibians.

Terrestrial Habitats.   Many toads and frogs, newts and salamanders spend most of their time in the terrestrial habitat. One must be familiar with the distribution and habitat requirement of such animals before collecting can be successful. In forested situations where salamanders are common, specimens may be found in and under rotten logs, bark, leaf litter, underground burrows, roadside seeps, in rock slides, and so on. Collecting at night with a lantern will often reveal numerous salamanders that are up for either mating or feeding. During the very heavy spring rains and early fall rains numerous frogs and salamanders can be delivered by driving the roads at night in a likely habitat. Blacktop roads with little traffic are the best for collecting. Start collecting soon after dark during the height of heavy pelting rainstorms. Drive slowly (between 10 and 15 miles an hour), using low beams from your automobile for light. You will quickly develop an "eye" that can distinguish very small salamanders from earthworms that are seen on the road. You will recognize that some areas are used as migratory paths by amphibians. Such runways should be frequented throughout the year in order to obtain a complete pattern of migration.

TRANSPORTING
LIVE SPECIMENS

It is often necessary to transport or ship live specimens over long distances. When working by automobile, prepare the specimens as described above (p. 227). The primary requirement is that the specimens be kept cool, moist, and clean. In warm, dry climates amphibians can easily be kept alive by placing their containers in the new, large, and inexpensive camping iceboxes. In extremely high temperatures small quantities of ice will keep the insides of such compartments relatively cool.

When shipping live specimens, air freight is the most desirable. Place specimens in a large stout plastic bag with wet toweling or wet moss. Have the bag almost completely inflated with air, twist the neck shut and seal the bag with rubber bands. Next, place the plastic bag in a large pasteboard box with a thick layer of loosely crumpled newspapers surrounding the bag; this will serve as a source of insulation and will keep the specimen from being damaged. Mark such containers "Living specimens for biological research," and "Keep away from extremes of high and low temperature." Styrofoam boxes and wet cloth bags also work well for shipment.

PRESERVATION TECHNIQUES

Procedure and Problems

Normally the process following collection includes (1) narcotizing, (2) killing, (3) positioning, (4) fixing, and (5) transfer to the final storage solution. Many workers do not wish to be bothered with this elaborate procedure while field collecting and simply drop living specimens into a proper preservative. However, the extra time involved in preservation is well worth the trouble, since the specimens are better looking and are more easily measured and stored. One problem that must be overcome in preservation is that of keeping the field data along with the specimens. This may mean that narcotizing and fixing will have to be conducted in numerous small containers to prevent the mixing of specimens. Sooner or later, a label must be attached to each specimen to keep its identity separate.

Narcotizing and/or Killing

Amphibians are allowed to swim in solutions containing narcotizing agents. These are usually local anesthetics which cause the specimens to lose consciousness or even die. It renders the specimen well relaxed and, therefore, permits it to be positioned in a desired manner.

Chloretone.   This is the best narcotizing agent for amphibians (see Appendix D). Make a stock saturated solution, filter, and add four parts of water to this for narcotizing. Specimens should be placed in containers with sufficient Chloretone solution to completely cover them. The animals will swim for several minutes before sinking to the bottom of the container. If the Chloretone solution is too strong, specimens will tend to contract and develop "kinks" in the tail and body. When the solution is properly diluted, however, specimens will die in a completely relaxed manner. An alternative is to narcotize specimens in a 5-percent alcohol solution.

Clove Oil.   The author finds that clove oil, which contains a phenol compound, is quite useful in narcotizing amphibians. Put 2 or 3 drops of clove oil in ⅔ quart of water, cap, and shake well. Use clove oil in the same manner as Chloretone. Specimens will usually sink to the bottom within 5 minutes and will be totally senseless in 10 minutes or so. A longer time is required for larger specimens.

Positioning and Fixing

This process involves placing specimens in pans of 8- to 10-percent formalin for from 48 hours to 1 week. Specimens should be thoroughly narcotized before transfer. Position salamanders with the tail out straight, as shown in Fig. 17–1A. If specimens are so long that they will not readily fit into containers, tails should be bent up along the side of the body, as shown in Fig. 17–1C. If necessary, use wax-bottom trays for fixing and pin the specimen in the desired position. Figure 17–1B shows the frog position that can most easily be worked with.

Heavy-bodied frogs and toads and salamanders should be injected in the body cavity with formalin or cut along the right side of the body, as shown in Fig. 17–1B. This permits the rapid preservation of the gut content, which otherwise would start to decay. Salamanders with a body diameter equal to or less than the diameter of a lead pencil need no incision or injection. Likewise, small frogs need no injection. If frogs tend to float in the fixing solution, the body and lungs are filled with air. Gently squeeze the abdomen, working from the posterior end up toward the anterior end, until the air is expelled from the body. Very large frogs and salamanders may require injection of formalin into the heavy muscles of the legs as well as in the abdomen.

Final Preservation

The following solutions are quite satisfactory for long-term preservation: 5-percent formalin, 70-percent ethyl alcohol, or 40- to 50-percent isopropyl alcohol. Specimens to be stored in formalin may be transferred directly from the fixing solution. Conversely, workers using alcohol generally prefer to soak out the formalin before transfer. To remove the formalin place the specimens in jars of fresh water for 24 hours or more. Change the water each day. Do not permit the specimens to become too soft, but rather transfer them to alcohol when most of the formalin has left the body. The alcohol should be changed after 24 hours to ensure that the fluids contained in the specimens have not diluted the preservative below a safe level.

Labeling and Records

Two types of labels may be used for amphibians, in addition to normal museum-bottle labels. These are attached directly to the specimen, either around the waist or around the hind appendage, as shown in Fig. 17–1, A and B. The most common label measures about ¼ inch by ¾ inch, and is made of a stout waterproof paper. It is tied to the specimen by means of a lightweight string or heavy button thread. This small label usually contains only the field collecting number or the museum acquisition number. Most museums log all new specimens in an acquisition book, along with the date and locality of collection and the collecting or field number. Every new acquisition has a new number and, therefore, specimens so numbered can easily be matched up with their field data. The second kind of label applied directly to specimens is somewhat larger, measuring approximately ½ inch by 2 inches. This label contains all of the pertinent field data, as shown. The larger label is much preferred for small collections where field notes are frequently separated from the collections. Two-hole labels are recommended, tied obliquely to salamanders for ease in groove counts.

Storage Methods

Although color cannot be preserved well in amphibians, every effort should be made to store specimens so as to retain as much color as possible. Therefore, specimens should be placed in museum jars, along with their data, tightly sealed, and stored in a darkened room or cabinet. Under ideal conditions, the temperature should not fluctuate in the storage room. With extreme temperature fluctuation the preservative expands and contracts and may eventually break the seal of the jar lid. Museum specimens should be checked at least twice a year and preservatives added when needed.

SHIPPING
FIELD SPECIMENS

When large numbers of specimens have been amassed on distant field trips it is simplest to ship them in bulk rather than in separate jars or other containers. Furthermore, specimens are usually held in formalin for shipment and are not soaked out and represerved in the final solution until they reach the home museum. Wrap individual specimens or groups of small specimens, together with their field data, in cheesecloth. Pour fresh preservative over the cheesecloth to keep the specimens moist. Next, place wrapped specimens in plastic bags, expel most of the air, twist the neck of the bag, and tie the bag closed with string. Numerous plastic bags of such specimens may now be packed in large metal drums which can be tightly sealed. No additional preservative is needed, so long as the plastic bag and drum are tightly sealed. If specimens do not completely fill all of the space within the drum, additional cloth or light paper toweling should be packed around the specimens to prevent any movement. Metal drums are, in turn, packed in wooden crates or in stout cardboard boxes. With the latter, pack a thick layer of crushed newspapers around the metal container.

An alternate method is that of sealing the specimens inside a series of plastic bags. Next, place the bags in a cardboard box and completely surround them with crushed newspaper. This technique will greatly reduce the weight and is more satisfactory for air freight. Specimens may remain in such containers for several months without damage. However, unpack specimens as soon as possible, to ensure their good condition.

SKELETAL TECHNIQUES

The skeleton is of extreme importance in amphibian taxonomy on all levels. Large animals may be fleshed out, dried, and cleaned by dermestid beetles. Most specimens, however, are too small for this and must be stained and cleared. Rare specimens which cannot be spared for "making skeletons" can be studied very adequately by means of stereoscopic X ray. This process reproduces the skeleton in such a way as to make it easily as valuable as a cleared and stained specimen. Because the procedures used for working with the skeletons of all vertebrates are so similar, these techniques will be discussed collectively in Chapter 21.

REFERENCES

Bishop, S., 1967, *Handbook of Salamanders,* Comstock, Ithaca, N.Y.

Cochran, D. M., 1961, *Living Amphibians of the World,* Doubleday, Garden City, New York.

Conant, R., 1958, *A Field Guide to Reptiles and Amphibians of the United States and Canada East of the 100 Meridan,* Houghton Mifflin, Boston, The Riverside Press, Cambridge.

Kincaid, T., 1948, To Preserve the Color Pattern of the Skin in Frogs, *Turtox News* **26**(2):50–51.

Knudsen, J. W., 1966, *Biological Techniques: Collecting, Preserving, and Illustrating Plants and Animals,* Harper & Row, New York.

Legler, J. M., 1964, Tape Recordings of Frog Calls, *Turtox News* **42**(2):68–69.

Ley, W., 1955, *Salamanders and Other Wonders,* Viking, New York.

Logier, E. B. S., 1952, *The Frogs, Toads, and Salamanders of Eastern Canada,* Clark, Irwin, Toronto.

Netting, G., and G. Orton, 1950, *A Field Guide to the Amphibia and Reptiles,* Houghton Mifflin, Boston.

Noble, G. K., 1954, *The Biology of the Amphibia,* Dover, New York.

Oliver, J., 1955, *Natural History of North American Amphibians and Reptiles,* Van Nostrand Reinhold, New York.

Schmidt, K. P., 1953, *A Checklist of North American Amphibians and Reptiles,* American Society of Ecthyology and Herpetology, Northridge, Calif.

Smith, M. A., 1951, *The British Amphibians and Reptiles,* Collins, London.

Stebbins, Robert C., 1954, *Amphibians and Reptiles of Western North America,* McGraw-Hill, New York.

Stebbins, R. C., 1966, *A Field Guide to Western Reptiles and Amphibians,* Houghton Mifflin, Boston, The Riverside Press, Cambridge.

Wright, A. H., and A. A. Wright, 1949, *Handbook of Frogs and Toads,* Comstock, Ithaca, N.Y.

Zim, H. S., and H. M. Smith, 1953, *Reptiles and Amphibians,* Western, Racine, Wis.

CHAPTER 18

The Reptiles

This diverse group is perhaps the most exciting and least understood of all vertebrates. Sadly, reptiles and amphibians are being pushed to extinction both specifically and geographically. Private or academic noncollecting will not help matters as long as wholesale capture and slaughter by "live" animal dealers is tolerated. Limited collecting is urged, therefore, unless specimens are required for research purposes.

FACTORS TO CONSIDER IN COLLECTING

A collector should advise himself of his state laws concerning the collection of reptiles. Most crocodilians and some turtles are protected by law generally throughout the United States. The Gila monster is protected throughout its range in the southwestern United States. Some states (such as California) protect all plants and animals; thus, a permit must be obtained before any collecting can be done. Because of the poor reputation of some snakes, such as the rattlers, "look-alike" species that are perfectly harmless are often killed by citizens who feel they are doing a good deed. Field collectors are often overly enthusiastic and may frequently take more specimens than a local population can afford. Teachers and collectors alike, therefore, should be highly conscious of the need for conservation and protection of reptilians and other animals to prevent them from becoming scarce or extinct.

Because reptiles are cold-blooded (poikilothermic) their metabolic rate is governed by environmental temperatures. Collectors can take advantage of this by keeping specimens relatively cool. Under such conditions snakes and lizards will require little food over long periods of time. Conversely, collectors should realize that desert reptilians are still very sensitive to extremely high temperatures. Reptiles utilize shade during the heat of the day and may suffer sunstroke if kept in the sun. When translated, these facts suggest that specimens should not be kept in closed automobiles that will overheat. Preferably, the windows should be open, the automobile should be in shade, or the specimens should be placed beneath the automobile while collectors are in the field.

Finally, body measurements are important taxonomic tools. These measurements (see Fig. 18–3, p. 242) are difficult to take on poorly preserved specimens. Therefore, collecting trips should be organized well enough so that specimens may either be properly preserved in the field or transported back to the laboratory for preservation.

FIELD DATA

The field notebook is perhaps more important than any other field tool, for unless "on the spot" records are kept, specimens will be of little value. Records should include the data and geographic locality (see Chapter 1) and as much physiographic and ecological data as possible. The plant formation, the terrain, air and soil temperatures, the microhabitat, and so on, should be described. In addition, details of any observed activity, or lack of activity, should be noted. For example, notes concerning mating, feeding, sunning, or other activities will all help to create a general description of the specimen's natural history. Color notes (see Chapter 1) should also be taken when specimens are intended for taxonomic purposes.

COLLECTING METHODS

Equipment and Field Problems

Reptiles are placed in cloth bags when captured; flour sacks or seed bags are excellent for this. Use bags made of stout canvas for poisonous snakes or large nonpoisonous specimens. Provide sacks with a tie string (not a draw-string) to secure the bag. Snakes are especially adept at working through weak spots in the seam of a bag or through the neck of the bag, unless it is properly tied. Figure 18–1 shows one method of "bagging" a specimen. The specimen is dropped into the bag, whereupon the mouth of the bag is closed and the sack is spun around to twist the neck. This prevents speci-mens from escaping while the neck of the bag is being tied. Note (Fig. 18–1C) the neck of the bag is doubled over and double tied.

Large lizards and nonpoisonous snakes are placed in bags as follows: While holding the specimen in your right hand reach all the way to the bot-tom of the bag, grip the specimen with your left hand through the bag, re-lease your right hand while still holding the specimen with your left hand, and twist the neck of the bag closed. Tie the bag as shown in Fig. 18–1C. Professional rattlesnake collectors attach a collecting sack to a stout hoop and handle. The bag then resembles a butterfly net. Specimens are trans-ferred to this bag by means of a snake stick or hook stick. The bag is closed with a light rope tied in an overhand knot around the neck of the bag. Finally, the neck is double tied, as shown in Fig. 18–1. This technique greatly lessens the danger of being bitten.

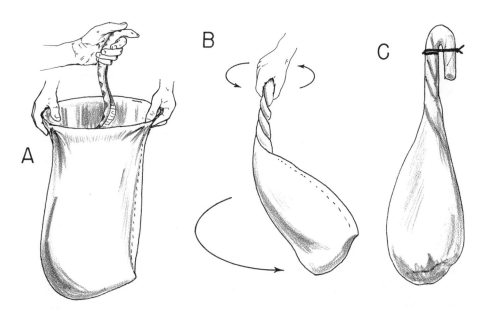

FIG. 18–1. Method of holding a snake and placing reptiles in a snake bag by first dropping the specimen in the bag, twisting the bag, and finally tying the neck.

Lizards

In the higher latitudes lizards remain in open country or brushy country, but are seldom found in heavy forests. Toward the tropics arboreal species are more numerous. Some legless lizards are found only in soft soil or sand where they burrow and hunt insect larvae for their food. Most lizards are extremely fast when sufficiently warm, and are usually alert enough to make approach difficult. Furthermore, lizards sun themselves near hiding places, into which they retreat when disturbed. Most lizards can autotomize (break off) their tails if handled roughly and, for this reason, must be captured in such a way as to prevent injury. Snares, .22-caliber bird shot shells, sling-shots, pry bars, and various other tools are used for capturing lizards.

The lizard snare is one of the best all-around collecting tools. Lizards are very tolerant of a snare and will permit you to try numerous times to slip the snare around their heads. Their usual reaction, if any, is to try to grab the snare as if it were a fly. To use the snare move very cautiously and quietly toward the lizard until it can be reached by the tip of the snare. Carefully work the snare over the lizard's head and then pull it closed. In closing the snare, always move the snare stick backward along the length of the lizard's body rather than forward and in front of the lizard. This will cause the snare to tighten around the neck just ahead of the shoulders rather than slipping off the lizard's head. Use 6-pound or 8-pound woven nylon fish line, or ½-pound to 1-pound monofilament nylon leader. Tie the snare and attach it to the snare rod, as shown in Fig. 18–2, A and B. Any long stick may serve as the snare rod, but a telescopic fishing pole is the most ideal.

Professional collectors use .22-caliber pistols with bird-shot shells, approach within 10 feet of a lizard specimen, and then shoot at it. A few shot generally stun (or, sometimes, kill) the specimen, giving the collector time to capture the lizard before it can escape. Specimens collected in this manner, however, must be preserved almost immediately to prevent spoilage. In the tropics many of the arboreal (tree-dwelling) lizards and snakes can be collected in no other way than by shooting. The old-fashioned slingshot is amazingly accurate (with a little practice) and quite useful for collecting lizards.

In desert areas snakes and lizards wander about at night a great deal during the early part of their breeding season. This usually occurs on those first few nights when the air temperature remains above 80° at least until after midnight. At such times drive slowly (between 15 and 20 miles an hour) along paved back roads in the desert, using the low beams of the car

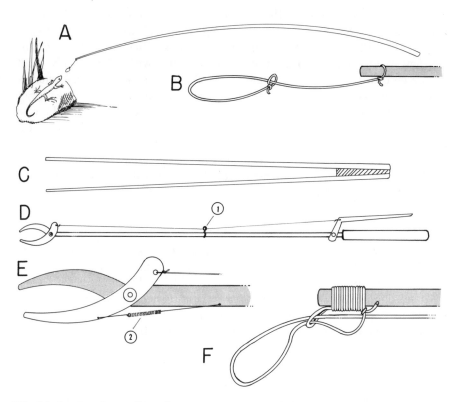

FIG. 18–2. Reptile-catching devices. A–B. Use and detail of the lizard snare. C. Wooden tongs for picking up poisonous reptiles. D–E. Details of a snake stick. (1) Guide, (2) spring. F. Details of a heavy-duty snare for capturing very large snakes.

for light. Specimens can be approached directly and picked up by hand under such conditions.

Snakes

Many of the techniques used for lizards are directly applicable to snakes. It is more difficult to predict where snakes will be found than where lizards may occur. Snakes that are coiled and sunning frequently give the collector sufficient time to snare them or to capture them in some other manner. However, many snakes will move quickly into heavy brush or seek out rodent burrows and must be captured before they can escape. The .22-caliber pistol with bird shot is quite good, especially for poisonous specimens. When you are equipped to do so, you can stop most of the little snakes by gently stepping on them, grasping them by the neck, and transferring them to a collecting bag. Some collectors use long wooden forceps (Fig. 18–2C) which are easily constructed in the field. A snake stick (Fig. 18–2, D and E) may be anything from a forked stick or a stick with a large screw hook in one end, to a stick with a single moveable jaw for the capture of snakes. Some of the collectors in Costa Rica used homemade snake sticks (Fig. 18–2). Note that the moveable jaw is kept open by means of a small coiled spring or a piece of surgical rubber tubing. The jaws are closed around the neck of the snake by pulling back on the string, either by means of a handle (Fig. 18–2D) or manually. Larger snakes in the open may be snared. Figure 18–2F shows how light, ⅛-inch sash cord can be tied to some convenient stick to form a temporary snare.

Turtles

Turtles present a special problem in museum work because of their unusual size and dimensions. It is often impossible to find museum containers that can hold larger specimens and, therefore, the collector should be somewhat selective when this is possible. Most land turtles can be picked up and placed in cloth bags. Aquatic specimens are easily caught with skin-diving equipment if the water is clear. The fish seine, described in Chapter 16, may also be used to good advantage in shallow ponds, lakes, and slow-moving rivers.

PRESERVATION TECHNIQUES

Snakes and Lizards

Procedure.   The procedure for proper preservation of reptiles is as follows: (1) killing, (2) positioning the specimens, (3) fixing specimens, (4) labeling, (5) final preservation and storage.

Killing Methods.   If the facilities are available, by far the best and simplest method for killing any reptile is freezing. Place specimens in cloth bags, along with their field data, put the specimens in the freezing compartment of the refrigerator overnight, remove, and thaw. Because reduced temperatures lower the metabolic rate of reptiles, the specimens simply go to "sleep" as they cool off and are thus killed in a very humane way. The only danger with this technique is that tails, toes, legs, or bodies may break should the specimens be dropped or mishandled before they are thawed. Once completely thawed, they are treated like any other specimen.

One widely used method of killing reptiles is that of injecting them with a 10-percent nembutal solution. Some workers use ether or chloroform in place of nembutal. The nembutal method is excellent for field work where limited facilities are available. The amount to be injected must be determined by the size of the specimen. Nembutal, being an anesthetic, renders the specimens unconscious. Because an overdose is used specimens quickly die and are ready for preservation. Be careful not to inject too much nembutal into small lizards and snakes for muscle contractions and kinking may occur.

Snakes and lizards are often drowned in warm water, but this is a slow process. It is true that hot water will rapidly kill reptiles but it also causes kinking and body contraction. One method for drowning reptiles is as follows: Place the specimen in a net bag (not a cloth bag) with a weight. Drop the specimen in a bucket of water which is warm to the touch but not uncomfortably hot. Another method is as follows: Place the specimen in a jar. Fill another container with about twice the amount of water required to fill the specimen jar. The water should be warm to the touch but not too hot. Quickly remove the lid and fill the jar to capacity, replace the lid and seal it tightly. If successful there will be very few or no air bubbles in the jar. The warm water accelerates the metabolic rate of reptiles and usually kills them by asphyxiation. Obviously, poisonous snakes are too dangerous to handle in this manner. If kinking occurs in specimens attempt to straighten the spinal column immediately before the muscles become set.

Positioning, Fixing, and Labeling.   Figure 18–3, A and B, shows the typical measurements recorded for lizards and snakes (total length, snout-to-vent length, tail length). The reptile skin inhibits preservatives from entering the body quickly enough to inhibit rotting. All small lizards should either be injected in the body cavity with formalin or should have a cut made on the left ventral side of the body. Larger lizards should also be injected in each leg segment and just underneath the skin at the base of the tail. If a hypodermic is not available, use a very sharp scalpel or razor blade and cut small slits in the limbs and tail. Caution must be used when working around the tail of a lizard since it can break off at any time. Snakes are either injected with 10-percent formalin every inch along the length of the body cavity, or else they are cut on the left ventral surface of the body. These cuts should measure 1 to 2 inches in length, should penetrate into the body cavity, and should be about 1 inch apart.

Two types of labels may be used (Fig. 18–3D). The typical label measures ¼ by ¾ inch and has either the collector's field number or the museum acquisition number. In either case, this number refers to a complete

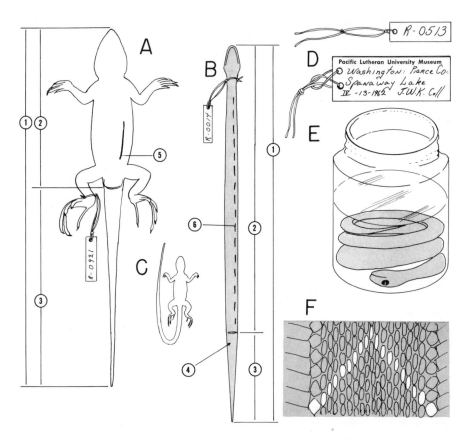

FIG. 18–3. Measuring, tagging, and positioning reptiles for preservation. A. General lizard techniques. B. General snake techniques. C. Posture for long-tailed lizards. D. Tags bearing the museum number or the complete field data. E. A snake coiled during the initial phase of preservation. F. Scale-count technique. Make the scale count following the normal scale pattern as shown by the white scales: the scale count of this specimen is 19. (1) Total length, (2) snout-to-vent length, (3) tail length, (4) inject or press here to evert the hemipenis, (5) cut or inject at this point, (6) make a series of cuts or injections along this line.

set of field data. The second label bears all of the field data on one side and the taxonomic information on the other. This label is preferred in small collections where a filing system or other method of recording data is not maintained. The label is attached around the neck of a snake or around the hind limb or waist of a lizard. Labels should be made of waterproof paper and tied to the specimens with lightweight, but strong, string.

When positioning specimens, always keep in mind the containers in which they must eventually be placed. Lizards are positioned in pans containing 10-percent formalin. Specimens that are short enough to fit in quart bottles have the tail extended; those which are extremely long have the tail brought up along the length of the body (Fig. 18–3, A and C, respectively). Snakes are coiled, belly up, in a bottle containing 10-percent formalin. This posture is essential in snakes that have been cut, to permit gases to escape. It is not essential for injected specimens, but it is, nevertheless, convenient. To coil a snake, hold the specimen by the tail, and by twisting the tail coil the specimen into the bottle. Finally, pour enough formalin to cover it by 1 or 2 inches.

Fixation requires from 48 hours to a week. Observe the specimens carefully for any discoloring of the ventral abdominal surface. Discoloring indicates that the specimen is rotting internally. Should this happen, the rotten area should be injected with 10-percent formalin and then cut with a razor blade.

Final Preservation and Storage.   Small and medium-sized reptiles may be kept in 6-percent formalin; large specimens should remain in 10-percent formalin. Because of formalin's very disagreeable nature, most workers prefer to transfer specimens from formalin to alcohol. Seventy-percent ethyl alcohol or 50-percent isopropyl alcohol is suitable. Specimens may be transferred directly from formalin to alcohol or soaked to remove the formalin, as follows: Place specimens in a jar of clear water and let stand for one or more days. Change the water daily. Watch carefully for a faint scum on the surface of the water, which indicates that the formalin level is so low that bacteria can begin to grow. Transfer the specimens to alcohol and replace the solution after 24 hours with a fresh solution. Seal specimens in airtight museum jars, along with their field data. Specimens should be kept in the dark to retard color loss. Check museum specimens at least twice a year and replace any fluid that has been lost as a result of evaporation.

Large Snakes and Lizards.   When specimens captured in the field are too large for standard museum containers they should be skinned and preserved. Treat snakes as follows: Record color notes and make the three standard measurements (Fig. 18–3). Next, make a slit down the ventral surface of the body from the neck to the vent. Cut the body in two at the base of the skull and at the base of the tail, being careful not to injure the skin. Next, peel the skin away from the main part of the body; roll up the head, tail, and skin; label; and preserve in 10-percent formalin.

Large lizards are treated as follows: Record color notes and make measurements (Fig. 18–3A). Make a slit on the ventral surface of the body from the neck to the vent. Begin peeling the skin away from the body and, when possible, sever the neck, the base of the limbs, and the tail from the main trunk. Now remove the main part of the body, but leave the skull, limbs, and tail intact. Do not attempt to skin the tail, as it may autotomize at any time. Label the specimen, roll it up, and preserve it in 10-percent formalin.

Turtles

Killing Methods. Any of the methods described for snakes and lizards may be used for turtles. Freezing is by far the easiest method; drowning in a mesh bag, injection with nembutal, and other methods may work as well.

Preserving Small Specimens. When the specimen is dead pull the head, tail, and limbs out of the shell so that they are exposed. Make slits between the neck and forelimbs and between the tail and hind limbs; inject formalin into the head, neck, and limbs. Place the specimen directly in a jar with 10-percent formalin. After several days this solution may be replaced by a 6- to 8-percent formalin solution or by alcohol, as described above for snakes and lizards. Attach a specimen label to either the neck or the hind limb.

Wet-Dry Specimens. Perhaps the only suitable way to handle really large turtles, such as marine turtles, other than having them stuffed by a taxidermist, is drying the shells and preserving the head, limbs and tail. To achieve this, separate the plastron (ventral shell) from the carapace (dorsal shell) by cutting the bridge that joins these two shells. This is usually cut with a hacksaw, but may be severed with a knife in marine turtles. Next, carefully cut the skin free from the shell so that the head, neck, and two anterior limbs may be removed as a single unit. If this unit is too large to preserve in one piece, sever the forelimbs from the neck. Do the same for the tail and two hind limbs. Carefully inject the limbs or make frequent small cuts to ensure proper preservation. Preserve these portions in 10-percent formalin, replacing the solution after 3 or 4 days. Use large glass containers or metal drums lined with plastic bags for these portions.

Next, remove all of the flesh from the two shells, label each shell individually, wash it with a solution of 10-percent formalin on the inner surfaces, and dry in a dark but airy place.

SHIPPING
FIELD SPECIMENS

All of the techniques described in Chapter 17 for the amphibians may be employed for shipping field specimens of reptiles. Such specimens are positioned, fixed, and stored in formalin until after they have been shipped.

SKELETAL TECHNIQUES

The bony structures of reptiles are perhaps as important in taxonomy as external characteristics. At any rate, taxonomists must prepare some specimens for skeletal studies. Small snakes and lizards are cleared and stained by standard, but somewhat modified, techniques. Larger specimens are skinned and then cleaned by dermestid beetles. Because of the similarity in techniques used for all vertebrates, skeletal techniques for reptiles will be discussed in Chapter 21.

REFERENCES

Carr, A., 1952, *Handbook of Turtles*, Comstock, Ithaca, N.Y.

Conant, R., 1958, *A Field Guide to Reptiles and Amphibians of the United States and Canada East of the 100th Meridian*, Houghton Mifflin, Boston, The Riverside Press, Cambridge.

Ditmars, R., 1936, *The Reptiles of North America*, Doubleday, Garden City, N.Y.

Ditmars, R., 1939, *Fieldbook of North American Snakes*, Doubleday, Garden City, N.Y.

Ditmars, R., 1966, *Reptiles of the World*, Macmillan, New York.

Knudsen, J. W., 1966, *Biological Techniques: Collecting, Preserving, and Illustrating Plants and Animals*, Harper & Row, New York.

Pope, Clifford, 1939, *Turtles of the United States and Canada*, Knopf, New York.

Pope, Clifford, 1955, *The Reptile World*, Knopf, New York.

Schmidt, K., and D. Davis, 1941, *Fieldbook of Snakes of North America and Canada*, Putnam, New York.

Schmidt, K. P., and R. F. Inger, 1958, *Living Reptiles of the World*, Doubleday, Garden City, N.Y.

Smith, H. M., 1946, *Handbook of Lizards*, Comstock, Ithaca, N.Y.

Stebbins, R. C., 1966, *A Field Guide to Western Reptiles and Amphibians*, Houghton Mifflin, Boston, The Riverside Press, Cambridge.

Wright, A. H., and A. A. Wright, 1957, *Handbook of Snakes*, Comstock, Ithaca, N.Y.

CHAPTER 19

The Birds

Happily, bird study today involves more field observation and identification than field collecting. The great popularity of bird study has supported the production of excellent field guides. This book treats the collection and preparation of specimens but, for the sake of length, leaves out advice on ectoparasites and crop contents of collected specimens, bird nests and egg collection, field study, and bird banding. Readers are directed to *Biological Techniques* (Knudsen, 1966) for further material.

COLLECTING TECHNIQUES

Laws and Permits for Collecting

One must assume that almost all birds are protected by federal law (if they migrate across state boundaries) and also by state law, and that a federal permit and state permit must be obtained before they can be legally possessed (this includes road kills) or collected. However, numerous migratory game birds and upland game birds may be taken in season with a state

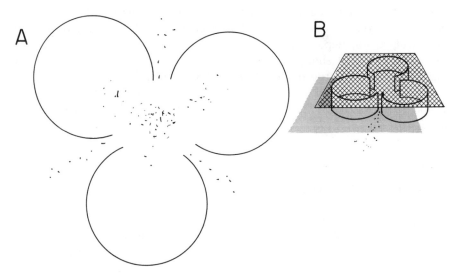

FIG. 19–1. The Potter bird trap. A. A diagrammatic view showing three partial circles of screen wire set in position to form three funnels, and bait (grain) scattered between the funnels. B. The Potter trap set and ready to work. Use hardware cloth construction, and weight top of trap with a stone.

hunting license. Hunting seasons, bag limits, and the like, are set each year by the state fish and game department. In addition, certain birds listed by state as "unprotected" may be hunted at any time of the year. For example, in Washington State a recent list of unprotected birds included such species as the crow, the English sparrow, and the starling. The collector must correspond with his state's fish and game department to ascertain the legality of collecting migratory birds in season and to ascertain those species which are unprotected.

The procedure for obtaining bird permits is as follows: First, the federal permit must be obtained; then, the state permit. Contact the nearest regional office of the U.S. Fish and Wildlife Service for application blanks for the federal permit. When this permit is obtained, application blanks will be available from the state fish and game department's main headquarters.

Shotguns and Shooting

Birds are most satisfactorily collected with a shotgun. Usually, no single gun serves as a good all-round collecting tool. Rather, guns of different sizes are required for different types of birds. The 12-gauge, 16-gauge, and 20-gauge shotguns are used for larger birds, with shot sizes ranging from

No. 9 to No. 2 shot, depending on the size of the specimen and the distance of the shot. Whether low-base or the more powerful high-base shells are used will, again, depend on the size of the specimen and the distance of the shot. The .410 shotgun is an intermediate gun which occasionally is quite effective for large birds and may also be used for smaller specimens. Serious collectors often modify pistols, such as the .38 special or the .45, by reaming out the barrel to remove the rifling. Pistol shells are reloaded by hand with shot of various sizes.

Care of Specimens

Once the bird has been collected, plan to spend adequate time in temporary field preparation. With a forceps plug all shot holes with cotton to prevent bleeding. Shot holes should be plugged immediately, even if there seems to be no local bleeding, because bleeding usually commences an hour or two after the specimen has been killed. Wash any blood from the feathers with cold clear water and blot the feathers dry. If blood appears in the mouth, hang the bird head down and permit the blood to drain. Then plug the throat with cotton to prevent further bleeding. Plug the anus also, if this seems necessary. Record field notes and then tag the specimen with its field number by means of a tie-on tag. Finally, wrap the specimen in absorbent paper, such as newspaper, and pack it in a canvas field bag. Small birds are usually wrapped in a cone of paper and then placed in the field bag. Specimens so treated will remain in good condition throughout the day's hunting. In warm weather it is essential to mount the specimen soon after collecting, or to pack it in an ice chest or refrigerator.

When it is not convenient to mount specimens immediately after collecting, simply place the wrapped field specimen in two plastic bags, each sealed in turn, and place this parcel in the freezer. Make sure the field data have been included with the specimen. The double plastic bag will prevent undue loss of moisture and thus preserve the specimen until the time of mounting.

Killing Injured Birds

One encounters birds from time to time that have broken wings or have been otherwise injured seriously. Such birds are quite defensive but nevertheless should be humanely killed to prevent further suffering. Killing is accomplished by depressing the chest and causing almost instant suffocation. Grasp smaller birds beneath the wings and firmly squeeze the chest for about one minute. Grasp larger birds, such as ducks, by the two wing tips while they are lying on the ground, and kill by depressing the chest with your foot.

Window Kills and D.O.R. Specimens

More and more birds are falling victim to large picture windows. They see the reflection of the sky in the window and apparently assume they can fly directly through the window. Thus, in any neighborhood a surprising number of birds will be killed in this fashion. Collectors should let it be known, among their faculties, neighbors, scout groups, and so on, that specimens will be welcomed. If you are always ready to retrieve specimens when notified, a long and unending supply will develop. The collector should always be prepared to care for specimens found dead on the road (D.O.R.). At certain times of the year an alarming number of hawks, owls, robins, and other birds are killed on the road. The best specimens are usually those which bounce from an automobile and land at the side of the road.

Skinning and Salting Field Specimens

On extensive field trips consider salting bird skins, shipping them to the laboratory, and making them up as specimens at a later date. If facilities are limited or the humidity such that skins are difficult to dry, the salting technique is often the only one for adequate field collecting.

George Hudson of Washington State University, who has prepared numerous skins in this manner, recommends the following techniques to his students: After the bird skin has been removed from the body (as directed below) carefully remove flesh and fat clinging to the inside of the skin. Turn the skin right side out and apply salt generously to all surfaces of the inside of the skin, filling the head and neck. Large birds with fleshy feet (such as hawks, ducks, and gulls) should have the tendons removed from the legs (as directed below) and salt worked down into the cavities occupied by the tendons. The present author prefers to inject the toes, feet, and tendon canals with 10-percent formalin in preference to cutting and salting. However, Hudson states that "in hot weather the skin should be loosened from the bone over the entire foot and tarsus and each toe should be cut open along the entire bottom surface. Particular care should be taken to loosen the skin at the upper end of the tarsus and to pack in much salt." Next, fill the skin full of salt, using up to 1 gallon or more for a goose, as the excess salt will soak up the moisture and help keep the feathers dry. The next day, dump the salt out of the skin and dry it (the salt) over a stove. Refill the skin with dry salt and repeat this daily until the skin is fairly dry but still flexible. Finally tie a label to the foot of the bird, shake out all excess salt, roll the skin, and seal it in a plastic bag to prevent further drying. Ship the specimen to the home laboratory by air mail, whereupon it may be stored in a freezer or cold room until it is put up as a round mount.

Field Data for Collected Specimens

The collector should keep a field notebook (see Chapter 1) wherein the collecting data are recorded. The date and geographical location should be clearly defined and notes should be taken to describe the habitat, weather conditions, behavior of the specimen, and so on. Color noting for birds is generally not essential in that feather colors are not altered if the specimen is properly cared for. However, the bills of ducks and geese and the feet of larger birds, such as sea gulls, ducks, geese, and coots, should be noted as to color, for these generally darken.

MOUNTING
AND PRESERVATION

Mounting Methods

Museum specimens are not prepared like the traditional taxidermy mount in which the bird is given a lifelike pose. Rather, museum mounts or round mounts are made of bird skins. Such mounts are designed to show all of the essential areas of the plumage and other morphological characteristics while keeping the bulk of the specimen to a minimum. Dozens of round mounts may be stored in the same space required for a single taxidermy mount. Very small birds (hummingbirds and the like) lend themselves well to the wet-and-dry technique which avoids the tedious problem of skinning and stuffing. Larger birds, such as the great blue heron, are frequently prepared as flat mounts to conserve space. The collector must decide which technique is most suitable for each specimen collected. However, the standard procedure is the round mount.

The Mounting Kit

An old fishing-tackle box makes an excellent kit. The following things should be provided: a scalpel or sharp knife, a carborundum sharpening stone, a pair of fine-pointed scissors and a coarse pair of scissors, a fine-pointed forceps and a coarse pair of forceps, a package of assorted sewing needles, a spool of No. 2 cotton or linen thread and a spool of heavy cotton or linen button thread (white), a package of common pins, a 30-centimeter ruler graduated in millimeters, a 10-cubic centimeter hypodermic with an assortment of coarse and fine needles, a pound of absorbent (long-fibered) cotton, 1 pound of powdered arsenic trioxide or Boraxo, a pound of coarse corn meal or hardwood sawdust, bird labels, dip pen and waterproof ink. Optional materials that may prove useful, especially if mammals are also to be mounted, are the side-cutting pliers and a pair of dividers. The cotton is used to make up "bird bodies" for stuffing; the corn meal is used to absorb grease and blood during the skinning process. Fine hardwood sawdust is used in the tanning industry as a fur-drying compound. The liberal use of corn meal or sawdust will often make the difference between a greasy, poorly mounted skin and a clean skin. Powdered arsenic trioxide is used to poison the skin, contract the skin around the feather, and, in general, cure it. Boraxo (pure borax may alter the color) is probably just as good as arsenic trioxide for curing skins and is preferable for student use.

Problems to Anticipate

Blood.   If there is any bleeding from the flesh of the body during the skinning process, immediately add large quantities of corn meal, soak up the blood, and remove the corn meal. Should blood get on the feathers, either from the incision or from shot holes, wash it away with cold clear water before it has a chance to dry. Wash individual feathers with a cotton swab, blot, and dry. If dried blood is found on the feathers, scrape this away from each individual feather with a pocketknife or fingernail. Brush each feather until it is clean. Usually, a feather which is too badly damaged by blood may be removed from the body without damaging the appearance of the skin.

Fat Skins and Grease.   Use corn meal generously when skinning birds that contain large quantities of fat. After the skin has been removed, turn the skin inside out and scrape the fat away from the skin with a dull pocketknife. Continuously work corn meal into the fatty areas to absorb the grease. A short length of hacksaw blade is useful to break up the fatty tissue in the feather tracks of ducks and similar birds. Work this tool vigorously along the skin, but be careful not to break the skin. Next, scrape with a pocketknife and, finally, clean with several applications of corn meal.

The following technique is used by some museums when dealing with greasy skins. Most of the grease is removed as described above. The skin is then washed in warm water and detergent, particular attention being given the fatty areas. When the fat has been sufficiently removed from the skin, most of the water is squeezed from the skin and feathers. Next, a drying compound is liberally added to the skin and feathers. A product known as "Blue Cloud" (the Don Company, Gardena, California: a bath product made for drying live chinchillas) is used as a drying agent by many professional collectors. Corn meal or hardwood sawdust may also be used. When the moisture has been absorbed from the feathers, a small automobile vacuum cleaner is used to retrieve the drying compound. An air hose could be substituted to blow the compound out of the feathers. Finally, the feathers are fluffed up and the skin is then ready for stuffing.

Determining Sex.   The sex of the bird must always be noted on its label, along with size and relative development of the gonad. Some birds are sexually dimorphic and present no problem, but where the sexes are identical in external appearance, one must dissect the body to locate the gonad. As soon as the body has been removed from the skin, cut through the body wall on the left side of the rib cage. Open the body cavity and carefully move the intestines to one side. The gonads are located anteriorly to the kidneys, the male having two testes while the female has but a single ovary on the left side. Be careful not to mistake the yellowish adrenal glands, also associated with the kidneys, for the gonads. The testes are oval, whereas the ovary is irregularly lobed. Record the sex, make a diagram of the actual size of the gonad, and measure the gonad for the field notebook. Include the sex and the gonad drawing on the bird label.

Skinning Techniques.   This discussion deals with the general techniques used for all birds. In special instances additional techniques may be required for some particular species. These are: (1) techniques for large-headed birds such as ducks, geese, and woodpeckers, where the neck skin is too small in diameter to be stretched over the large head; and (2) techniques for the wings and feet of large birds where considerable muscle and other tissue must be removed. These techniques are described below.

1. The primary incision.   Spread several sheets of newspaper on a table, set out your dissecting equipment, and pour out a quantity of corn meal in a wide-mouth container so that it is readily accessible. Have additional newspapers at hand to replace the original working surface should it become soiled with blood and grease. The skinning process should not be done so quickly that feathers are lost, that the skin is ripped or stretched, and so on. Nevertheless, work as quickly as possible to prevent the skin from becoming too dry once it is partially removed from the carcass. Now, place the bird on its back with its head away from you. The ventral abdominal feathers are attached to the skin in two rows or tracks, one on either side of the ventral line. With the butt of the scalpel push the feathers away from the midventral line, beginning at the middle of the sternum and moving back toward the vent. On smaller birds these feathers can be blown to one side quite efficiently. With your fingers or a pair of forceps grasp the skin on the ventral line, half way between the vent and sternum, pull the skin away from the thin abdominal muscles, and snip the skin with a pair of sharp-pointed scissors. Next, insert the scissors through the cut of the skin and work forward to the middle of the sternum and then backward to the vent.

2. The knees.   Grasp the skin on one side of the incision and gently loosen the skin, working toward one side. As soon as possible, add some corn meal to absorb any grease or body juices. If the primary incision cuts through the muscle wall into the body cavity, plug this with corn meal and cotton. Be constantly on guard to keep your hands as clean as possible. Now grasp the bird's leg by the heel and push the leg forward so that the knee will protrude into view (Fig. 19–2A). Separate the tibio-tarsus at the joint where it joins the femur with a scissors, and sever the remaining muscle tissue as indicated in Fig. 19–2A. Add corn meal to the cut flesh to absorb any blood. Repeat this process for the other knee.

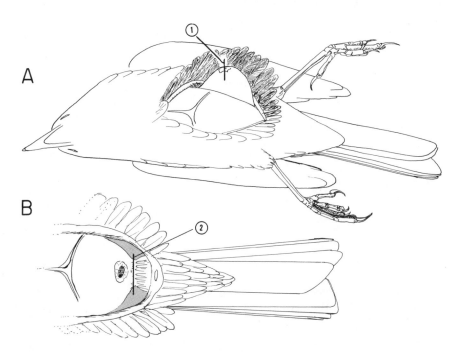

FIG. 19–2.   Initial steps in mounting birds. See text for complete instructions. (1) Cut the knee at this point, (2) carefully separate the body from the tail at this point.

3. The tail.   With a scissors cut around the terminal end of the digestive tract, plug or replug the digestive tract with cotton, and carefully loosen the skin around the body, back toward the tail. Exercise extreme care in this part of the skinning job, as the skin on the sides and back of the rump is very thin, especially in smaller birds, and is readily torn. This area of skin seems subject to extreme drying, which also facilitates tearing. In most round mounts this particular area of skin is all that holds the tail onto the body and, thus, it must be as strong as possible. In larger birds the skin will often loosen all the way around the body, even across the back of the rump. In smaller birds, however, the skin adheres very tightly to the tail. The object is to cut the tail free from the body just ahead of the tail feathers. Exercise extreme care not to cut the basal ends of the tail feathers, as this may cause them to fall out. With a fine-pointed scissors cut minute bits of flesh until the tail is severed, as indicated in Fig. 19–2B. If rips are made in the skin, these may be sewn at the time.

4. The body and neck.   Now use your fingers or the butt of the scalpel to carefully push the skin away from the body, working forward toward the shoulders and neck. When you reach the base of the wings separate the humerus bones from their sockets and continue gently loosening this skin around the shoulders. The skin of smaller birds is loosely attached in this area, but is usually more firmly attached in larger birds. Again, care must be taken not to rip the skin. Continue to use corn meal to pick up grease and blood.

By this time the skin is almost turned inside out. It would seem that the feathers around the neck area would be badly damaged by this procedure, but no damage is done. Using your fingers to push the skin forward, continue to turn the skin inside out until you reach the base of the skull. As noted above, if the bird you are skinning is a large-headed specimen, it will have to be treated in a special manner (described below).

5. Skinning the skull.   Using your fingers (Fig. 19–3A), work the skin up over the skull until the thin, funnel-shaped pieces of skin protruding into the ear become visible (Fig. 19–3B). The thin head skin will dry very rapidly and may need periodic moistening with clean water to prevent tearing. Now, do not cut off the "ear skin" where it joins the head skin, but rather pull it out of the auditory canal with your fingernail or a pair of blunt forceps. The skin will now slip over the top of the head and, as it is drawn forward, the eyes will be exposed. On the first dissection the student is usually amazed at the extremely large size of the eye as compared to the relatively small opening. The eyelids attach to the eyeball by means of fine connective tissue. Use extreme care in detaching the eyelids from the eye. Some workers use a very fine-pointed scissors for this procedure. The author,

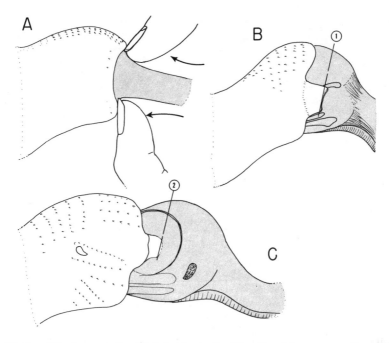

FIG. 19–3. Skinning out the bird head. A. Pushing the skin up over the skull. B. Skinning around the ear. C. Skinning around the eye. (1) Cut the ear close to the skull, (2) begin cutting the eye at this point (see text).

however, prefers to place a scalpel with the cutting edge down on the eye, as shown in Fig. 19–3C (note, the scalpel is not placed on the skin). The scalpel is drawn back and forth in a sawing motion and simultaneously is moved slowly forward directly over the pupil of the eye. During this procedure the head skin is pulled gently forward to facilitate detaching of the eyelids. After both eyelids are free, the skin needs to be worked forward only a short distance, so that it is parallel with the anterior margin of the eye socket.

6. Cleaning the skull.   Loosen the eyes where they join the skull, reach in with a forceps and grasp the optic nerves, and remove the eyeballs from the skull, taking care not to puncture them as the fluid may soil the feathers. Next, pull the tongue out of the mouth by cutting where necessary. Push the points of a scissors up into the head and cut across the roof of the mouth on the ventral side of the eye sockets (Fig. 19–4B). Next, cut the floor of the brain case, starting at the base of the tongue and working posteriorly toward the base of the skull (Fig. 19–4A). Continue these cuts up the sides and across the top of the back of the skull (Fig. 19–4C). Now remove the

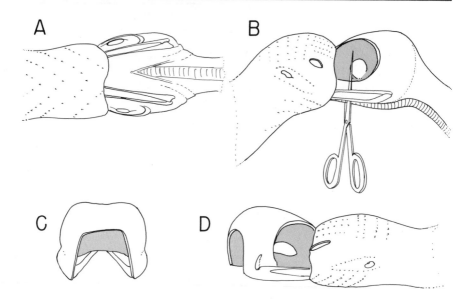

FIG. 19–4.   Cleaning the bird skull. See text for complete instructions.

neck and body from the skull, pick away the remaining portions of the brain, and remove any large quantities of flesh that adhere to the skull. At this point, moisten the skin which covers the skull with clear water. Dust the skin and the entire skull with arsenic or Boraxo. Keep a small piece of cotton in the arsenic jar which can be picked up by forceps and used for dusting the skin. Make two large cotton wads, each equal to the size of the eyes, and insert one in each orbit. With your fingers slowly work the skin back over the head while it is still moist. The skin must be pushed posteriorly at least until it reaches the back margin of the orbits. At this point the bill may be grasped where it protrudes through the neck skin, and the entire head and neck may be turned right side out. Never attempt this if the neck and head skin is too dry.

7. The wings and feet.   Working from the inner side of the empty skin, remove the flesh from the humerus, radius, and ulna bones in the wings. Dust these bones with arsenic or Boraxo and turn the wings right side out. Next, remove the flesh from the tibio-tarsus, dust the skin and the bone with arsenic or Boraxo, and wrap the bone with enough cotton to replace the removed flesh. By pulling the feet, turn the legs right side out.

8. Treating the skin.   Finally, treat the entire inner surface of the skin with arsenic or Boraxo. Next, roll the skin up and temporarily place it in a clean plastic bag to prevent it from drying out. If the skin seems exceptionally dry, moisten it slightly with clean water before adding the arsenic. Set the skin aside while preparing an artificial body.

Large-Headed Birds.   When dealing with large-headed birds, we deviate from the normal skinning procedure, described above, by severing the neck (not the skin) about one-third its length back from the base of the skull. Next, an incision is made on the median-dorsal part of the skin over the base of the skull (Fig. 19–5A). The cut should be just large enough to permit the head to pass through it. The head skin is now turned inside out and worked over the skull, as described above.

Don Pattie (in personal communication) recommends making the incision on the ventral part of the neck at the posterior end of the head (Fig. 19–5B). This technique requires a smaller opening, is easier to sew, and is less conspicuous. You have the option with this technique of not skinning out the entire head, inasmuch as the tongue, brain, large muscles, and even

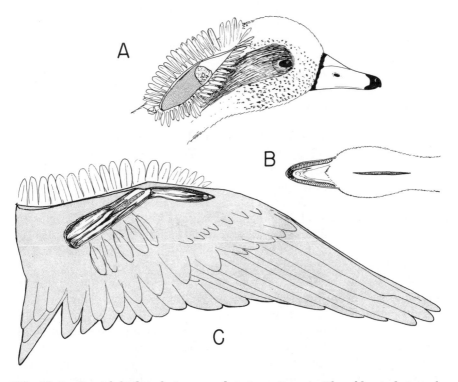

FIG. 19–5. Special bird techniques used in mounting. A. The older technique for working with large-headed birds, see text. B. The Pattie technique for working with large-headed birds, see text. C. Removing the musculature from large bird wings.

the eyes can be removed by working up from the ventral surface of the skull.

Wings and Feet of Large Birds.   Larger birds such as owls, hawks, sea gulls, ducks, and the like, must have the muscle tissue removed from the length of the wing. After the entire carcass has been removed from the skin, pin out one wing exposing the ventral surface. Make an incision between the radius and ulna bones, and continue this on down between the meta-carpal bones (forearm and hand, Fig. 19–5C). Move the skin away and remove the major muscles and sever them at both the origin and insertion. Dust the entire area with arsenic or Boraxo and close the incision with one or two stitches. Remove the flesh from the humerus bone, working from the inside of the skin as directed above.

Many workers prefer to remove the tendons where they extend down into the feet parallel to the tarso-metatarsal bones. This is achieved by cutting the tendons at the distal end of the tibio-tarsus while this bone is being fleshed. Next, an incision is made on the underside of the foot just at the base of the toes. Through this incision the tendons may be hooked with a wire probe and pulled down out of the leg. The author prefers simply to inject the foot and tendon canal with 10-percent formalin which will pre-serve this tissue during the drying process and will eliminate the necessity for removing the tendons.

Stuffing.   Make an artificial body which consists of a central stick wrapped with cotton (Fig. 19–6, A or B). The central stick may be made of hardwood dowel, a piece of wooden box, or any appropriate material. The length of this stick is equal to the length of the body, neck, and part of the head (Fig. 19–6C). Some workers prefer to leave an additional length protruding from the anal opening of the skin (Fig. 19–6B). The crossed feet, and even the tail, may be secured to this additional length of stick in the finished specimen, if desired. Cut a blunt point on the anterior end of the stick. Unroll a few feet of long-fibered, absorbent cotton. Such a roll is usually about 1 inch thick and 10 inches wide, with fibers running length-wise. The trick of making the body is to pull off long, wide, but very thin layers of cotton from the roll and wrap these around the stick. Note that the bulk of the body should be equal to that of the carcass, whereas the bulk of the artificial neck is slightly greater than the real neck.

For large birds the procedure of cutting the stick and building up cotton for the neck portion is quite similar. However, excelsior or other material may be substituted for the cotton body. Tie small quantities of excelsior to the stick and slowly build up the artificial body to the size of the actual

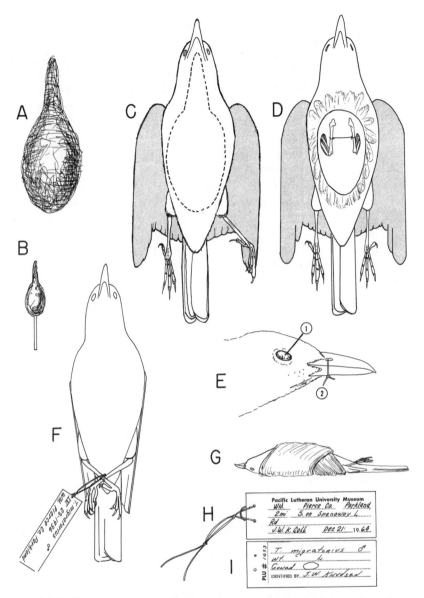

FIG. 19–6. Completing the bird mount. A–B. Two types of artificial bodies (see text). C. The position of the artificial body shown by dotted lines. D. Tying the humeral bones. E. Techniques of finishing the bird head. (1) Pull cotton part way out through the eye opening, (2) tie the bill closed. F. The finished position with the label attached to the crossed legs. G. The specimen temporarily wrapped in cotton for drying. H–I. Two sides of a standard bird tag.

carcass. Continuously wrap the excelsior with string which must be just tight enough to hold it in place during stuffing.

Now set the completed artificial body aside and unroll the bird skin. Attach a string to one humerus just above the elbow, and tie the other end to the opposite humerus (Fig. 19–6D). The distance between humeri varies with the size of the bird. The object of tying the humeri is to prevent the wings from drooping away from the body and to give the round mount a more natural appearance.

Next, insert the tip of the artificial body well up into the skull of the bird and work the skin down around the body. You are now ready to sew up the ventral incision. With a needle sew a thread through one side of the anterior opening. Do not tie a knot in the thread but, rather, knot the thread to the skin itself at this point. Now stitch through the opposite side a little lower down, cross back over and stitch through the skin toward the middle of the incision, and repeat the process, going from side to side and crossing back and forth four or five times. Do not draw the thread up until it has been stitched the entire length of the incision. Then, starting at the anterior end, pull the skin closed over the artificial body, knotting the thread at the last stitch. It is not serious if the skin will not completely close over the body, as the feathers will cover the incision. Tie the bill closed by passing a needle and thread through the nostrils, from side to side, and knotting the thread underneath the mandible (Fig. 19–6E). Cross the feet and tie these together with the label string (Fig. 19–6F), attaching the feet to the protruding stick, if so desired.

The bird is now ready for grooming and wrapping. While the skin is still moist the feathers can easily be arranged in a natural position. Begin at the head end and work down, stroking the feathers with your fingers or a soft brush. Arrange the wings, check the details of the rump feathers, and, in general, "put the feathers back together." Finally, the round mount is wrapped in long, wide, but thin wisps of cotton or in pieces of jeweler's cotton (Fig. 19–6G). This holds the feathers in place during the drying of the skin. The cotton fibers will adhere to each other and hold firmly. Birds with large heads, crests, or other plumage should have the head turned to one side to display these features. The bird should now be placed in a dust-free, warm, and dry place for a few days to several weeks until the skin is thoroughly dry. Some workers suggest unwrapping the bird periodically during the first day of drying and rearranging the feathers if they become disarranged by the shrinking of the skin. This is usually not necessary if extreme care was taken in the initial operation.

Labeling.   Each collector should adopt a standard label made of a good grade of high-rag-content paper. The information is variously arranged on such labels. One side usually has the sex of the bird, the locality and date of collection, and the collector's name. The other side of the label contains the scientific name of the bird, a description of the gonad, and may include such things as weight and length (Fig. 19–6, H and I, respectively).

Wet and Dry Mounts

Small birds, such as hummingbirds, kinglets, and creepers, are quite difficult for the beginner to skin and mount. Small birds can safely be preserved with 10-percent formalin and then dried and placed in museum trays along with other skins.

Using a fine hypodermic needle inject 10-percent formalin into the body cavity and head (in the throat region). Next, prepare a small foot label and attach this to the legs, as described above. Carefully groom the feathers and position the body and then wrap the specimen in thin wisps of cotton, as described above. When it is completely wrapped, weight the body by adding a small fishing weight with some additional cotton. Submerge the specimen in 10-percent formalin for 24 hours, then remove the specimen and set it aside to dry. The weight may be removed from the specimen, but do not remove the remainder of the wrapping which holds the feathers. Remove the wrapping after the specimen is thoroughly dry. Specimens so prepared need no special care, but are kept with regular bird skins and are fumigated periodically with paradichlorobenzene.

Flat Skins. Owing to restrictions in space and other factors, larger birds are occasionally prepared as flat skins. Specimens are skinned, cleaned, and treated with arsenic or Boraxo, as described above. After this the feathers are groomed and the skin is allowed to dry in a flat position. The skin is not split and spread out as is a mammal pelt, but rather resembles the skin in Fig. 19–6D at the time it is flattened. Birds with long necks should have the head brought alongside the body and the feet directed forward if they protrude too far posteriorly. Such skins can be relaxed at any time and made up into reasonably good round mounts. To relax dried skins place them in a humidifier equipped with chlorocresol (see Chapter 12 for directions).

Skeletal Techniques

On many occasions bird skeletons are needed for taxonomic purposes or for classroom teaching. Very small birds or chicks are best cleared and stained; larger birds should be cleaned by dermestid beetles or by some other technique. Because the skeletal techniques for all vertebrates are so similar, the reader is referred to the single treatment of this topic in Chapter 21.

Storage of Skins

There are three general requirements for storing bird or mammal skins. Storage cabinets must be kept in a dry room; a damp, unheated room is not satisfactory. Second, the storage cabinet must be light-proof so that the colors of the specimens will not be altered. Finally, storage cabinets must be insect-proof and must be fumigated periodically with paradichloroben-zene. The design and construction of a suitable storage cabinet will be found in the textbook, *Biological Techniques* (Knudsen, 1966).

Measuring Bird Skins

Unlike mammals, which must be measured before mounting, birds' measurements may all be adequately taken from the prepared skin. With the exception of weight and the possible exception of total length, the measurements of prepared round mounts are very suitable for scientific work. Thus, many bird collectors omit measurement at the time of mounting. The standard measurements are shown in Fig. 19–7; they are the bill length, the total length, the body length, the tail length, the wing length, and the "tarsal" length (tarso-metatarsus).

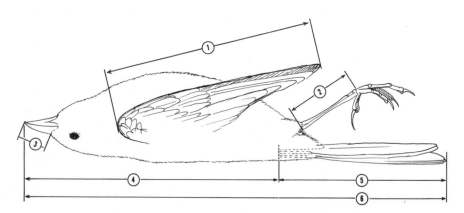

FIG. 19–7. Some standard bird measurements. (1) Wing length, (2) tarsal length, (3) bill length, (4) body length measured to base of tail feathers, (5) tail length, (6) total length.

REFERENCES

Allen, A. A., 1930, *The Book of Bird Life,* Van Nostrand Reinhold, N.Y.

Campbell, Bruce, 1953, *Finding Nests,* Collins, London.

Fisher, J., and R. M. Lockley, 1954, *Sea Birds,* Houghton Mifflin, Boston.

Knudsen, J. W., 1966, *Biological Techniques: Collecting, Preserving, and Illustrating Plants and Animals,* Harper & Row, N.Y.

Murphy, R. C., 1936, *The Oceanic Birds of South America,* Macmillan, N.Y.

Peterson, R. T., 1947, *A Field Guide to the Birds,* Houghton Mifflin, Boston.

Peterson, R. T., 1961, *A Field Guide to the Western Birds,* Houghton Mifflin, Boston.

Peterson, R. T., 1962, *How to Know the Birds,* Houghton Mifflin, Boston.

Peterson, R. T., G. Mountfort, and P. Hollom, 1966, *A Field Guide to the Birds of Britain and Europe,* Houghton Mifflin, Boston.

Pettingill, O., 1951, *A Guide to Bird Finding East of the Mississippi,* Oxford Univ. Press, N.Y.

Pettingill, O., 1953, *A Guide to Bird Finding West of the Mississippi,* Oxford Univ. Press, N.Y.

Pettingill, O., 1970, *Ornithology in Lab and Field,* Burgess, Minneapolis.

Pough, R. H., 1951, *Audubon Land Bird Guide: Small Land Birds of Eastern and Central North America from Southern Texas to Central Greenland,* Doubleday, Garden City, N.Y.

Pough, R. H., 1951, *Audubon Water Bird Guide: Water, Game, and Large Land Birds,* Doubleday, Garden City, N.Y.

Pough, R. H., 1957, *Audubon Western Bird Guide,* Doubleday, Garden City, N.Y.

Ridgway, R., 1891, Directions for Collecting Birds, *Bull. U.S. Nat. Mus.* 39(A): 5–27.

Scott, P., 1968, *A Coloured Key to the Wildfowl of the World,* Heinman, N.Y.

Tinbergen, N., 1961, *The Herring Gull's World: A Study of the Social Behaviour of Birds,* The New Naturalist Series No. 9, Basic Books, N.Y. (Available in paperback from Harper & Row, 1971.)

CHAPTER 20

The Mammals

The mammals of natural areas may be studied in the field by day, with field glasses and a telephoto camera; or by night by using a blind and a light made from an automobile battery and headlight (covered with red cellophane, which nocturnal mammals cannot see). Mammal studies will require some collecting of live specimens, but even here it should be conservative. Highway kills offer an excellent opportunty to preserve valuable specimens. To prepare these for study and then donate them to a natural history museum together with field notes would be a great service. In addition, new methods must be devised whereby mammals can be studied and valuable data obtained without the collection of specimens. Small rodents provide the best source of live-catch material for student use, as their breeding habits are geared to heavy predation.

COLLECTING TECHNIQUES

The small mammals can readily be collected once their natural history is understood. Like all other creatures, mammals are "found only where you find them." As poor as this statement seems, it points up the problem that most beginning collectors have in locating small or large mammals. It is surprising how numerous some of the small rodents are, on the one hand, and how few of us are really aware of their presence, on the other hand. Among the requisites for locating mammals is a knowledge of mammalian natural history, periods of behavior, habitat requirements and hiding-place requirements, food requirements and the distribution of food organisms, and, finally, knowledge of the telltale signs such as runways, droppings, and the like, which mammals leave as evidence of their presence. One of the excellent books dealing with this last-named topic is *A Field Guide to Animal Tracks* by Murie (1954). Among the many field guides dealing with mammal recognition, distribution, habitat, and signs are the following: Anthony (1942), Booth (1971), Burt and Grossenheider (1964), Cameron (1956), Palmer (1954).

Laws and Permits

A few mammals, such as the fur seal and the sea otter, are totally protected by law and the possession of any specimens of these species is illegal without a federal permit. A larger number of mammals is also classified by each state as fur-bearing mammals which may be taken in season with a trapper's permit. Write to your state fish and game department for full details on obtaining such a permit and on protected animals.

Trapping

Snap Traps. The standard mouse and rat trap sold everywhere is excellent for collecting mice, rats, chipmunks, squirrels, weasels, shrews, and so on. Because of its smaller size, the mouse trap tends to damage many skulls, whereas the rat trap occasionally tears the skin of specimens. The Museum Special Rodent Trap (Animal Trap Company, Lititz, Pa.) is intermediate in size between the mouse and rat traps and has a "softer" spring. Thus, this trap (Fig. 20–1A) is a favorite of mammal collectors in that the bail seldom catches a specimen across the skull or crushes its body.

Many kinds of baits may be tried, but those containing peanut butter seem to work the best. A mixture of rolled oats, water, and peanut butter should be sticky enough to adhere to the trigger of the trap. Addition of chopped raisins and chopped nuts to the peanut butter-water-oat bait seldom improves the catch per unit of bait. At least, the author finds that he eats more of this bait than he uses on the traps. Plain cheese or plain bacon seldom work as well as the peanut-butter mixtures.

Each trap should be provided with a stout string about 20 inches long,

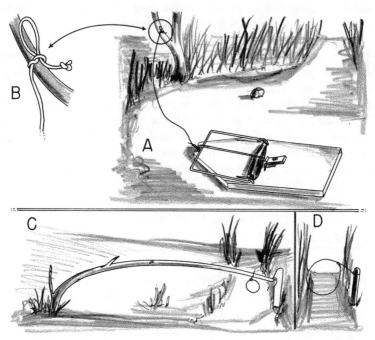

FIG. 20–1. Small mammal traps. A. Museum special set in runway. B. A handy knot for securing snap traps. C. A small snare set in a rodent runway. D. A wire snare set in a rodent runway.

so that it may be anchored to local vegetation. The trapper should also carry a piece of red or white cloth to make flag markers for relocating the traps. There are possibly as many ways of setting traps as there are trappers. Where there is a natural orientation of the vegetation in the area to be trapped, such as along a stream bed or path, the trapper may simply move along at random, setting one or two traps by each large clump of brush. The trap may be anchored quickly, using the combination overhand and slipknot (Fig. 20–1B). A strip of cloth should be tied to some brush in a conspicuous place near each set, in order to guide you back to your traps in the morning.

In open country, such as prairie grassland or desert and semidesert where the vegetation is sparse and lacks any natural formation, traps may be set every ten paces or at any likely looking site. When working in the open this way, relocate your traps by using cloth flags, by following a natural path and scratching a deep mark into the ground by each trap, or by dragging a heavy stick along the ground to make a furrow which will denote your trapping route the next morning.

Traps should be picked up very early in the morning, as specimens left in the sunshine soon begin to rot and may be attacked by ants. In the desert the author generally runs the trapline twice during the night and again the first thing in the morning, as both predators and the warm evening temperatures are effective in reducing the specimens. When it is impossible to mount the specimens immediately, place them in a plastic bag, along with their data, and put this in a camp freezer chest with dry ice. So preserved, the specimens may be handled at your leisure.

Weasels and some rats are attracted to fresh, bloody meat. For weasels, nail the trap onto a fence post or other object near a weasel runway, with the trigger between 6 and 8 inches off the ground. When the weasel stands up to investigate the bloody-meat bait, the bail of the trap will swing downward, catching the animal across the shoulders, and will kill it instantly.

When you are interested in ectoparasites the specimens must be isolated —at least, according to species and, preferably, by individuals. Fleas and lice are slow to leave their host animals and, thus, some of them will be present at the time of pickup. Therefore, if individual specimens are placed in small cloth or paper bags at the time of pickup the relationship between ectoparasite and host will be maintained. Locate the ectoparasites as described in Chapters 12 and 14.

Steel Traps.   Steel foot traps consist of two heavy jaws which catch the mammal by the foot when it steps on the trigger. These traps are cruel, to say the least, and should not be used in random settings. It is desirable to select the largest-size foot trap designated for a particular kind of mammal, rather than a smaller size, in that the larger trap will grip the specimen higher up on the leg and thus prevent it from twisting or chewing its foot off. Needless to say, steel traps should be tended faithfully in the interest of the animal caught, or not used at all.

The kind of bait used with steel traps, if any, will depend on the type of animal to be caught. For example, wolves, coyotes, and foxes may be trapped at natural signposts (places where these animals habitually urinate) or baited with the urine from a female dog which is in heat, with the carcass of some animal, or with a stink bait. A stink bait is prepared by sealing fish or meat in a bottle half-filled with water. After two or three weeks' rotting this fluid makes an excellent lure. One of the many ways of using the stink bait is as follows: Dig a shallow hole large enough to accommodate the trap and its stake and chain, sprinkle the trap with the stink bait, set the trap and place it in the hole, with the pan about 2 inches from the surface of the ground. Next, carefully pack dirt around the outside of the jaws and over the springs, place some cotton between the jaws and under the pan, and put a piece of light tissue paper over the jaws to prevent dirt from getting underneath the trigger or pan. Finally, sprinkle dirt over the entire trap so as to conceal it. The set is complete when all traces of the trapper have been removed and an additional drop of the stink bait is placed over the trap site and on some of the surrounding vegetation. The addition of an unwashed tin can near the trap site may help the catch. The animal is caught when it attempts to dig up what it thinks is a buried portion of rotten meat.

Many traps are placed in such a fashion as to take advantage of the normal movements of the animals sought. For example, wild cats and wild dogs, as well as other large mammals, usually step over small limbs, logs, or stones that are present in their runways. Thus, the steel trap may be set in the place where the mammal is most likely to step when crossing such an object.

Commercial Live-Catch Traps.   A number of commercially made live-catch traps, such as the Havahart trap and Sherman trap, are very humane and are thus recommended. The Sherman Company produces small, light-weight, all-metal mouse and rat traps which are collapsible and, therefore, take up very little room in a field pack. This is an important factor to consider when ordering live-catch traps.

Can and Bottle Traps.   Many small rodents and shrews can readily be trapped by burying cans or bottles flush with the ground and then suspending an appropriate bait (peanut butter or meat, respectively) over the opening. These animals will simply wander into such traps and will find it impossible to escape. Since shrews are carnivorous and usually eat one another, most workers partially fill the can or bottle with water and thus drown the specimens as they are caught.

Trapping Bats.   Bats are usually obtained by shooting, netting, or by mist nets similar to those used for birds. (There is a discussion of the use and sources of mist nets in the parent textbook, *Biological Techniques,* Knudsen, 1966.) Mist nets are set for bats in places where these animals fly, such as between rows of trees, across pathways, and along or over small streams. Fruit-eating bats are more easily taken than insect-eating bats, since the latter use their "radar" more than the former. Likewise, on clear moonlit nights bats tend to navigate as much by vision as by radar, and thus fail to detect the net. Conversely, on very dark nights many bats will fly directly to the net, then turn and go over it or under it since they have detected it by means of their radar.

Dalquest (1954), one of the pioneers at using the mist net for bats, has published an account of his collecting in tropical Mexico. His article should be consulted as it gives methods and procedures for collecting bats in great detail.

Many species of bats may be collected by stretching a taut wire or dark-colored string about 4 inches above the surface of slow-moving streams or ponds. Bats fly low over such bodies of water at dusk in order to drink or to catch insects. Therefore, they will strike the wire and fall into the water. This renders them helpless and will permit you to retrieve them with a dip net.

Gopher and Mole Traps. Most burrowing rodents, such as ground squirrels, can be shot or taken in foot traps placed directly outside their burrows. Gophers and moles, however, present a special problem and must be taken with traps designed especially for them. The wire gopher trap should be set 5 or 6 inches inside the mouth of a gopher burrow after it has been opened and cleaned. The hole is then partially plugged with vegetation or soil. The gopher is trapped when it pushes soil down the runway in an attempt to shut out the small amount of daylight showing.

Mole traps, on the other hand, are set in the middle of mole runways which are filled with soft, rock-free dirt. The mole is trapped when it noses its way through the soft dirt while moving along its runway.

Shooting Specimens

Little need be said about hunting techniques or equipment. It should be stated, however, that specimens should be shot with the least amount of damage possible. Head shots must be avoided, since they destroy the skull. When you are close to some specimen to be collected with a shotgun, aim beside the specimen rather than directly at it so that the animal will be killed with a minimum number of pellets.

As soon as a specimen has been shot, plug all of the shot holes with wads of absorbent cotton to prevent bleeding. If any blood occurs in the mouth of the animal, hang it up by its hind feet and permit this blood to drain out. Then, plug the throat with a large wad of cotton to prevent further bleeding. If it seems necessary, also plug the anus with cotton. Wash off all blood with fresh, cool water before it has a chance to set. Wrap the specimen in absorbent paper toweling or old newspaper or place the specimen in a small canvas sack. Specimens should be mounted as quickly as possible or frozen. However, it is best to wait at least one-half hour after the specimen has been shot, since there will be less bleeding in the mounting process.

Killing Injured Specimens

Trapped or wounded specimens must be dispatched immediately. Bear in mind that the skull cannot be injured, as this is needed for taxonomic work. Therefore, the best procedure for smaller mammals is to kill them by forcing the air out of the lungs. Place your foot on the chest of a foot-trapped specimen and hold it for a moment, or else place the specimen in a canvas bag and depress the lungs between your fingers or with your foot. Larger, dangerous mammals will have to be shot or drowned but should suffer the least amount of injury possible.

Salting Field Skins

Although smaller mammals may be prepared in the field, large specimens are usually salted, shipped to the laboratory, and then tanned commercially. Spread out the hide, fur side down, and cover it with a thick layer of ice cream salt or table salt. This salt will draw a lot of moisture from the atmosphere and from the skin itself, part of which may be drawn off. If necessary, scrape off the wet salt and dry it in an oven, then resalt the hide. When the hide is almost dry, shake off the excess salt, roll the hide fur side in, and pack it in a canvas or plastic bag. Be sure that a waterproof label has been attached to the hide giving all of the field data, which are also recorded in your field notebook.

MOUNTING
AND PRESERVATION

Choice of Technique

The standard method of preservation for mammals is that of the round mount for small mammals and the tanned pelt for larger specimens. The taxidermy mount, which poses the animal in a lifelike way, is not used in museum collections (except in public displays), since each specimen requires much more space for storage.

The case-skin mount, or flat-pelt technique, is very useful for small rodents when there is little time for standard mounting methods. Any small mammal intended for muscle study should be preserved in liquid. It is now preferable to preserve bats in liquid, although these may also be prepared as round mounts. Very small mammals such as shrews can be preserved in liquid and dried, using the "wet and dry mount" technique by modifying and adapting to mammals the method outlined in Chapter 19. Regardless of the technique, however, all mammals must be measured in a standard way to ensure an accurate record.

Measurements and Other Data

Figure 20–2 demonstrates the procedure used for measuring mammals. These measurements are the total length, the tail length, the foot length, and the ear length (Fig. 20–2, A, B, C, and D, respectively). All measurements are recorded in millimeters. Take the total length measurement by placing the specimen on a ruler, as shown, and measuring from the tip of the snout to the end of the vertebral column (not to the top of the hair on the tail). The tail length is again concerned with the length of the vertebral column (not the hair), and is measured from that point where the tail bends away from the body to the tip. The foot length is taken on the hind foot and

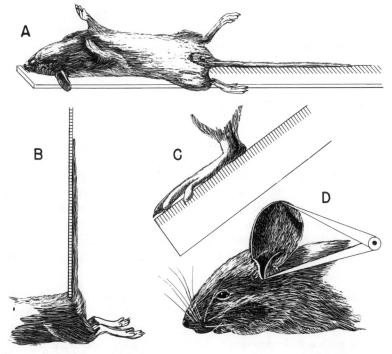

FIG. 20–2. Standard mammal measurements. A. Total length. B. Tail length. C. Foot length, measuring the hind foot. D. Measuring the ear: this may be done with a ruler, divider, or caliper.

may be somewhat variable depending on how worn the toenails are. A ruler (as shown) or a calipers (preferably) may be used for this measurement. The ear length, which can also be measured with a ruler, calipers, or dividers, is taken from the notch of the ear to the tip of the ear. Sometimes a second measurement is made from the crown of the skull to the tip of the ear. The four major measurements are also recorded in the field notebook and on the specimen label.

The collector should be concerned with a number of other data. The weight should be recorded in grams, as accurately as possible. Note the size and position of the gonads, as well as other data associated with reproduction such as pregnancy, the size of the embryo, the size of the mammary glands, whether or not lactation is occurring, and so on.

The Mounting Kit

The student is referred to the previous chapter for those tools indispensable in the mounting kit. In addition, provide yourself with an assortment of stainless steel or galvanized wires including, 16-, 18-, 20-, and 22-gauge sizes. For large mammals, 12-gauge or 14-gauge wire may be needed. In addition, a 30- or 50-cubic centimeter hypodermic syringe with a long needle is necessary for removing brains.

Problems to be Anticipated

Fat Skins.   During certain seasons of the year mammal skins contain large quantities of fat which must be removed. Fat left on the skin will eventually work through the pelt, discolor the fur, and may even cause the fur to slip. Once the skin has been removed from the carcass all small bits of meat and fat must be scraped away from it. Some workers wash fatty skins in warm water and detergent, paying particular attention to the concentrations of fat on the skin. When the fat has been extracted the skin is then squeezed, to remove most of the moisture, and dried. Copious amounts of corn meal or fine hardwood sawdust will quickly remove most of the moisture.

Bloodstains.   Blood should not be allowed to dry on the fur, but should be removed with clean cool water. Should some blood harden, however, mechanically dislodge it using your fingernail and a brush. Following this, clear water with a little detergent may be used with good results.

Determining Sex.   The external genitalia usually are diagnostic for sex determination. Some mammals are difficult to sex if they are immature, if they are not in a breeding condition at the time of mounting, or if the openings for the penis or vagina are similar in appearance. Many beginning students label female deer mice (*Peromyscus*) as males because the clitoris in some species is as long as or longer than the penis, thus causing confusion. Ideally, a dissection should be made of the carcass to observe and measure the gonads.

Food Samples.   Always examine the cheek pouches of small rodents and preserve or identify the contents. For a complete record, dissect the stomach and determine the type of food present. Record these data in the field notebook.

The Round Mount

Skinning.   This account deals directly with the standard procedure used for small mammals. Some special problems encountered with certain species will also be dealt with; these include methods of dealing with the muscular tails of beavers, muskrats, and porcupines, and methods of dealing with fleshy feet of raccoons, beavers, and others.

Perform the skinning job on old newspaper. Replace these should they become soiled or greasy. Provide a large quantity of corn meal in a wide-mouth container so that it will always be handy. Finally, set up those tools that you anticipate using so that they will be within reach.

1. The incision.   Weigh and measure the specimen, determine the sex, and record these data in your field notebook. Place the specimen on its back with its tail toward you, push the fur away from the midventral line of the body, grasp the belly skin between your fingers so as to pull it away from the body wall, and snip through the skin with a scissors. Now, slip the tip of the scissors underneath the skin and cut forward to the base of the sternum and then cut backward to the vent. If you have cut into the body cavity by mistake, plug the opening with cotton and add large quantities of corn meal to absorb any grease and blood. Following this, grasp the skin on one side of the incision and work it away from the body wall in the direction of the hind limb. On small rodents the skin is easily removed by simply pushing the body away from the skin with your fingers. In larger mammals, however, a great deal of connective tissue attaches the skin to the body and this must be broken with the butt end of a scalpel or very carefully cut. Two general dangers must be recognized here: (a) the skin may be torn or cut and (b) the skin may become stretched, making a "size 14" out of a "size 9" animal. Thus undue pulling should be avoided.

When the upper part of the leg is exposed, grasp the foot and push the leg forward, bending the knee as shown in Fig. 20–3B. Now work the skin down the lower part of the leg and sever the leg at the ankle. Pull the leg skin right side out. Repeat the process on the opposite leg.

At this point it should be noted that there is considerable difference in the technique for handling the lower limbs. Some workers leave the tibia and fibula of the lower leg in the finished skin by simply removing the tissue, adding arsenic, and wrapping the bones with cotton to replace the muscle tissue. Preferably, however, leg wires (discussed below) should be used in addition to these two bones or in place of them. In the former of these two methods, a wire is thrust into the foot all the way out to the end of the longest toe, and then is cut off even with the proximal end of the

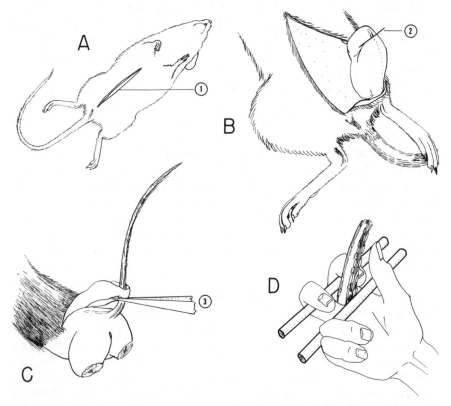

FIG. 20–3. Initial steps in skinning a small mammal. A. The primary incision (1). B. Severing the knee (2). C. Pulling the tail out of its skin with the aid of a pair of forceps (see text). D. A diagrammatic drawing showing how the tail of a large mammal is held for skinning (see text).

tibia bone. The bone is then poisoned with arsenic and the wire and bones are wrapped with cotton. The author prefers this last method, removing the tibia and fibula and replacing them with a leg wire and cotton. Nevertheless, the decision must be made as to which procedure will be used, for it involves removing the bones or leaving them intact.

2. The tail. After the knees have been worked out, cut the posterior end of the digestive tract away from the skin and plug the anal opening with cotton. Loosen the skin back around the body toward the base of the tail. In mice and other small mammals this is quickly accomplished and the tail is readily removed from the skin. Figure 20–3C shows the typical procedure for removing a mouse skin. Actually, the fingernails of the thumb and first finger are better substituted for forceps, but either method is satisfactory. Work the skin partly up the tail, not by pulling it, but by pushing it with your fingernails. As soon as there is room, place the blades of the forceps on either side of the tail (not the tail skin but the vertebral column) and by pulling the tail downward slip the tail out of the tail skin. The important thing here is that the forceps or your fingernails do not grip the skin and thus hold it tightly to the tail but rather are placed under the skin and thus push the skin up along the shank of the tail.

The tails of foxes, coyotes, and other large animals can be slipped in the same manner. The tail skin of any animal is most easily removed, however, soon after the death of the animal and may become very difficult with specimens that have been frozen for some time. To get a better grip on the tail of a large mammal cut two green sticks about ⅜ inch in diameter, place one on either side of the tail, and grip these with your fingers (Fig. 20–3D). Again, these are placed against the tail proper, not the tail skin, and are used to force the skin up the shank of the tail. If the skin refuses to slip, one must resort to cutting the tail skin open along the entire length of the ventral side to remove the vertebral column.

3. The body and neck. With the tail slipped from the tail skin, begin to turn the skin inside out and work it down over the length of the body. This can be done in a matter of seconds with small rodents, but care should be taken not to stretch the skin unduly. Sever the forelimbs at the elbows for ease in skinning, and then work the skin down each forelimb and sever the limb at the wrist. Continue to loosen the skin toward the head. About the head, neck, and shoulders connective tissues may attach the skin to the body. The skin of many rodents, rabbits, pikas, and other mammals will readily tear at these points and should be given additional care.

4. The head. Work the skin up over the base of the skull, loosening the connective tissue as you go. As the skin is worked toward the greatest diameter of the skull the funnel-like ear skin will be seen (Fig. 20–4A). Cut this with a fine-pointed scissors or scalpel where it enters the skull. If the

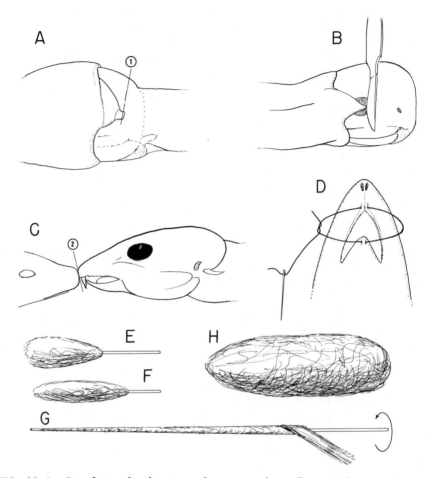

FIG. 20–4. Completing the skinning and preparing the stuffing. A. The ear technique. B. Working with the eye. C. Severing the nose. (1) Point of ear cut, (2) point of nose cut. D. Stitching up the mouth. E–F. Leg wires with cotton applied. G. Twisting the tail wire and applying cotton. H. The artificial body made of cotton. See text for complete instructions.

ear skin is cut out beyond this point a large gaping hole will appear in the finished mount. After the ears have been freed, work the skin down over the skull until the eyes come into view. A great deal of care is needed in cutting the eyelids free from the eye itself. The author prefers to place the blade of a scalpel down on the eyeball immediately behind the skin and to free the skin by sawing the scalpel back and forth across the pupil while slowly moving the scalpel in an anterior direction (Fig. 20–4B). The eye is seldom ruptured by this procedure and the eyelid is almost never cut. The alternative procedure is to use a fine-pointed scissors and carefully separate the eyelid from the eye. Finally, work the skin forward, separating it along the margins of the mouth and across the cartilaginous end of the nose (Fig. 20–4C).

5. Final preparation.   At this point all of the fat and muscle tissue clinging to the skin must be removed as described above. Next, the mouth is sewn closed while the skin is still turned inside out. The same procedure is used for small and large mammals. Pinch the skin of the lower jaw and pass a needle and thread through this, repeat this procedure with the skin of the upper jaw, and tie the mouth closed as shown in Fig. 20–4D. Now, if the skin is too dry moisten it slightly with water. Complete the preparation by dusting arsenic or Boraxo on the skin, roll it, and place it in a plastic bag to prevent drying during the preparation of the artificial body.

Problems with Tails.   Those animals that use their tails for swimming or defense, such as the beaver, muskrat or porcupine, are provided with a great deal of connective tissue that attaches the tail skin to the vertebral column along its entire length. Treat these animals as follows: Split the tail skin along its ventral surface from base to tip and carefully separate the connective tissue by means of a scalpel. When the time arrives, make an artificial tail out of a piece of straight-grained wood. The artificial tail must be as long as the true tail plus one-third of the body length. Add additional cotton if needed to fill out the space and sew the tail skin from tip to base with black thread.

Fleshy Feet.   The hind feet of some mammals such as raccoons, bears, and others, should be treated in one of two ways. Work the skin down over the ankle and out toward the toes far enough to remove the metatarsals and the associated tissue. The second method, which is quite useful for any mammal that may be used in a round mount, is injecting the feet and toes with 10-percent formalin to preserve them during the drying process. This method is quite simple and very satisfactory.

Stuffing.   The next task involves making an artificial body, leg wires, and a tail wire. Some practice may be needed in selecting the proper size of leg wire. This should be as large as possible and yet small enough to be worked part way up into the longest toe. The leg-wire length equals the length of the true leg, plus one third to one half of the body length. Cut all four leg wires to size. Now roll out a pound roll of long-fiber, absorbent cotton. Grasp a small quantity of this cotton from the surface of the roll and attempt to pull out as long a wisp as possible. The artificial leg is built up by wrapping a series of small thin wisps around the wire (Fig. 20–4, E and

F, hind limb and forelimb, respectively). The bare part of the wire must be long enough to extend through the sole or palm of the foot out into the longest toe. Make the cotton portion only as large as the leg itself, or slightly smaller than the leg itself. When applying cotton to wires for tails or legs the job may be simplified by moistening the wire so that the cotton will stick to it when first applied.

Cut the tail wire as long as the tail (measure from the carcass) and one third of the body. Pull some fine cotton wisps and begin wrapping these on the tail (Fig. 20–4G), beginning near the tip of the wire and working down toward the base.

The artificial body for small mammals is made by separating a long thin layer of cotton from the main roll. The width of this cotton should be slightly longer than the animal's carcass. Begin rolling the cotton until a roll equal to or very slightly larger than the diameter of the carcass is obtained. To develop the tapered head and rump contours, do the following: Grasp the roll firmly near one end, grab the tip of the roll at that same end, and pull the tip off and away from the main roll. When the main body is smoothed up again a tapered "head" will be produced. Follow the same procedure for the rump end, grasping the butt end of the roll and breaking the roll off at the required length. The butt end should be blunt (Fig. 20–4H).

Make sure that the skin has been poisoned and that the mouth has been sewn up. Turn the skin right side out and insert the artificial body. In small mammals this is easily accomplished by grasping the artificial body along its entire length with a 12-inch pair of forceps. The plug is then introduced into the skin and the forceps are removed. Next, insert the tail wire and leg wires, as shown in Fig. 20–5A. With a little practice you will learn to place the leg wires in position with no difficulty, as follows: Hold the foot firmly between the fingers of one hand, stretch the leg away from the body so as to straighten the skin, and insert the leg wire. With just a little probing the wire will find a channel to pass down through the leg and then should be directed into the longest toe. Finally, sew up the ventral incision, as follows: Run a needle and thread through one side of the incision on the anterior end. Tie the thread to the skin rather than using a knot in the thread. Stitch back and forth from side to side as shown in Fig. 20–5B, but do not draw the incision closed until you have traversed the entire length. Beginning at the anterior end, close the incision and tie the thread off at the posterior end.

The last task is attaching a foot label to the right hind limb, smoothing the fur, reproportioning the body, and pinning out the specimen to dry. A skull label which bears the same number as that used on the leg tag must also be attached to the skull at this time. If the skin has been stretched it will be somewhat baggy and should be "worked together and pushed together" so as to fit the artificial body as closely as possible. If the skin is too badly stretched the artificial body should be made somewhat larger than that of the carcass. Usually the skin will shrink during the drying process and will fit the artificial body very closely. The specimen is placed in a position that will take up the least amount of storage room. The forelimbs extend up along the chin and are pinned close to the body (Fig. 20–5D). The tail is pinned straight back from the body in line with the median axis. The hind feet are turned over (in an unnatural position) so that the fur side is up and the pads are turned down. These are pinned close to the tail for the reason that they would be easily broken if spread far apart.

After the tail and legs have been pinned in position, smooth the fur with

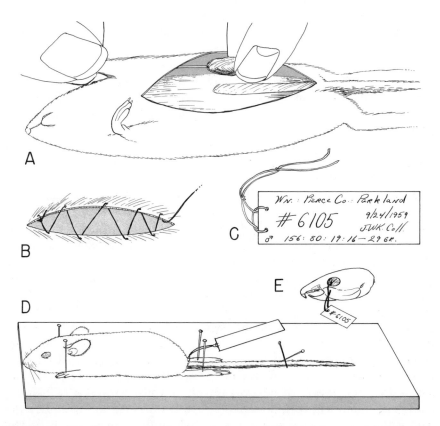

FIG. 20–5. The final mounting process. A. Inserting the foreleg wire. B. Stitching up the abdominal incision. C. A typical leg tag for a small mammal. Note the museum number and standard measurements given at the bottom. D. Pinning the specimen for drying. E. The partially clean skull with a label.

your finger or with a soft brush. The ears of rabbits should be folded back along the body and held together with a single stitch of thread. Some workers also fold back the ears of certain species of mice to protect the ears from damage. The author prefers to leave the ears sticking upward, since such specimens look more lifelike and are better as teaching aids for beginning classes. Finally, somewhere in this process, you should reach in through the eyelids and grasp the cotton of the artificial body and pull this out through the eyelid just far enough to keep the lids open. This is done to facilitate the study of colors around the eyelid. Place the specimen in a warm airy place to dry and protect it from direct sunlight or from flies which will deposit maggots on the skin.

Labeling.   Figure 20–5C shows a typical foot label. This contains the geographic locality and the date of collection, the name of the collector, and the standard measurements, which are always given in the order of total length, tail length, foot length, and ear length. The sex and weight of the specimen are also noted. Finally, the acquisition number or specimen number is given. Mammal collectors usually keep an accession notebook along with their field notebook in which each specimen is given a serial number. This number appears with the skin, skull, and skeleton. The same number is also entered in the museum records and thus represents the specimen in all of the records kept. The label is attached by means of a stout piece of cotton or linen button thread, tied as shown.

Skeleton or Skull.   The skull should always be labeled, cleaned, and saved for future studies. Many important taxonomic characteristics are concerned with the bony structure of the skull and the dentition, thus making the skull a valuable specimen. On the other hand, the skeleton is generally discarded unless it is needed for particular studies. Remove as much flesh from the skull as possible and attach a skull label bearing the accession number (Fig. 20–5E). Next, remove the tissue around the opening of the foramen magnum to expose the posterior end of the brain. Insert a long hypodermic needle through the foramen magnum and pass this through to the anterior part of the brain. Now inject water into the brain cavity in order to force the brain back out through the foramen magnum. If the skull is cracked this procedure will usually fail and the brain must be picked out with a dissecting needle. If the eyes are left in while the worker is "blowing the brains" there will be little or no danger of rupturing the skull. Dry the skull without any arsenic and in a place where flies cannot get at it. The skull is cleaned by dermestids at some later date. See Chapter 21 for details.

Case-Skin Mount. Anderson (1948) introduced a case-skin technique (in the first edition of his work, 1932) which is very useful when working with beginning students or when dealing with large numbers of specimens under field conditions. This technique may be used for mammals up to the size of small squirrels. In brief, the technique consists of skinning out the specimen as for a case pelt and attaching the skin to a permanent cardboard dryer, as shown in Fig. 20–6, A and B. The author uses a heavy grade of Bristol drawing board and carries three stock sizes in the field: 1⅛ by 12 inches, 2 by 18 inches, and 3 by 18 inches. These stock cardboards can be cut down and cut off to fit most small mammals as needed.

This technique can be modified in many ways. The author uses the following method, which is patterned fairly closely after Anderson's original technique: Make an incision across the body which cuts through the anal opening and extends part way out on the inner side of the hind limbs. Skin out the limbs, severing both feet at the ankles. Skin out the tail in the manner described for the round mount. Work the skin down over the back, severing the forelimbs, first at the elbows, and then at the wrists, and continue to remove the skin over the skull in the manner for the round mount. Next, sew up the mouth by the standard method, poison the skin with arsenic or Boraxo, and insert a tail wire which will extend halfway up into the body skin. Pull the skin over a cardboard stretcher which has been tapered to fit the head of the pelt. Attach the hind feet with a single loop of thread through the outside of the cardboard, and attach the tail with one or two loops of thread drawn through the cardboard stretcher as shown in Fig. 20–6. Some workers prefer leg wires, although these are not necessary

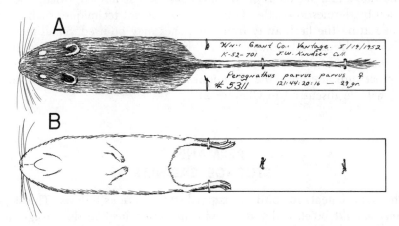

FIG. 20–6. The case-skin mount. A–B. Two sides of a rodent skin mounted on a cardboard stretcher. See text.

if the skin is given decent treatment in the museum. The forelimbs and ears should be folded back along the pelt. If necessary, slip the pelt into an envelope to hold the limbs and ears in position while drying. Record the standard field data, measurements, collector's name and date, and other data on the white surface of the Bristol board.

Liquid Preservation

Any mammal specimens required for muscle studies should be carefully preserved in formalin. Since so many of the taxonomic characteristics of bats are lost in the round mount, it has also become standard procedure to preserve most bat specimens in liquid. Specimens should be measured, weighed, examined for ectoparasites, and logged in the field or accession notebook. Next, inject the body cavity and larger areas of muscle with 10-percent formalin. Finally, wash the specimen with water and a little detergent to remove the grease from the body and fur. Washing is essential to permit rapid penetration of the body by the formalin. Be sure to use labels that will not disintegrate in liquid. Store specimens in airtight museum jars (filled with 10-percent formalin) in a dark place to prevent color breakdown in the more highly pigmented species.

SKELETAL TECHNIQUES

The skulls and skeletons of mammals are best if fleshed, dried, and then cleaned by dermestid beetles. Other less convenient techniques such as boiling or rotting the flesh away from the bones may be used. Small mammals (either adult or embryonic) should be cleared, stained, and stored in glycerin to prevent bone damage that would occur from the normal cleaning procedure. Because the techniques for all vertebrates are so similar, these will be discussed in Chapter 21.

PERMANENT
STORAGE CABINETS

The requirements for bird or mammal storage are as follows: The storage cabinet should be in a dry room with moderate heat to discourage mold. The cabinet itself must be light-proof and must be provided with a seal that will (1) prevent the entry of insect pests and (2) retard the loss of the paradichlorobenzene fumigant. Specimens are stored in paper-lined trays and arranged taxonomically. The mammal or bird case may be any size that is convenient, so long as the above requirements are met. See the parent textbook, *Biological Techniques* (Knudsen, 1966) for methods and diagrams for building storage cabinets.

REFERENCES

Anderson, R. M., 1948, Methods of Collecting and Preserving Vertebrate Animals, *Biol. Bull. No. 39,* pp. VII+164, Nat. Mus. Canada, Ottawa.

Anthony, H. E., 1942, *Mammals of America,* Garden City Books, New York.

Booth, E. S., 1971, *How to Know the Mammals,* Brown, Dubuque, Iowa.

Bourlière, F., 1955, *Mammals of the World,* Knopf, New York.

Burt, W., and R. Grossenheider, 1964, *A Field Guide to the Mammals,* Houghton Mifflin, Boston.

Cahalane, V. H., 1966, *Mammals of North America,* Macmillan, New York.

Cameron, Austin W., 1956, *A Guide to Eastern Canadian Mammals,* Natl. Mus. of Canada, Ottawa.

Dalquest, W. W., 1954, Netting Bats in Tropical Mexico, *Trans. Kan. Acad. Sci.* 57(1):1–10.

Glass, Bryan, 1951, *A Key to the Skulls of North American Mammals,* Burgess, Minneapolis.

Hall, E. R., and K. Kelson, 1959, *The Mammals of North America,* Ronald, New York.

Ingles, L. G., 1965, *Mammals of the Pacific States: California, Oregon, Washington,* Stanford Univ. Press, Stanford, Calif.

Knudsen, J. W., 1966, *Biological Techniques: Collecting, Preserving, and Illustrating Plants and Animals,* Harper & Row, New York.

Murie, Olaus, 1954, *A Field Guide to Animal Tracks,* Houghton Mifflin, Boston.

Palmer, R., 1954, *The Mammal Guide,* Doubleday, Garden City, N.Y.

Sanderson, I., 1955, *Living Mammals of the World,* Doubleday, Garden City, N.Y.

Young, J. Z., 1957, *The Life of Mammals,* Oxford Univ. Press, N.Y.

Zim, H., and D. Hoffmeister, 1955, *Mammals,* Western, Racine, Wis.

21

Vertebrate
Skeletal Techniques

Osteological material has long served as the basis for studies of vertebrate relationships, classification, and evolution. Obviously, osteological material presents one of the few means of comparing present-day animals with fossil specimens. The study of bone structures is becoming increasingly important, especially in the field of taxonomy on the generic, specific, and subspecific levels.

Skeletal specimens are prepared in a number of ways, depending on the intended use. Small vertebrates, including embryological specimens, can be studied conveniently by clearing and staining, or if rare, by means of X-ray techniques. Larger vertebrate skeletons are usually cleaned and dried, and may be stored as disarticulated or partially articulated skeletons for museum and taxonomic purposes or as fully articulated and mounted skeletons for teaching and demonstration. Cartilaginous skeletons, such as those of sharks and rays, require different techniques of cleaning and preservation for classroom use. For the latter (X-ray technique, mounting skeletons, and cartilaginous skeletal techniques), see *Biological Techniques* (Knudsen, 1966).

CLEARING
AND STAINING TECHNIQUES

Limitations and Uses

The clearing and staining technique, useful for small vertebrates, involves staining the osteological material and clearing all of the body tissues to make study possible (Fig. 21–1). As Davis and Gore (1947) point out, this method makes use of the Schultze (1897) technique of clearing tissue with KOH. They note that this technique is often only a supplement to other techniques. However, it has the advantage of preparing the entire skeleton with no loss or damage of small bones or distortion as a result of drying and shrinking of cartilaginous elements. Groups of bones, such as the girdles, are always intact and left in the proper spatial relationships to other bony elements. Therefore, whenever small vertebrates can be spared for "making skeleton," this technique should be one of the first used.

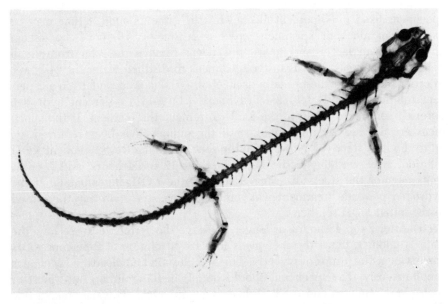

FIG. 21–1. A photograph of a specimen prepared by the clearing and staining technique, courtesy of Dr. David Wake. Note that the entire animal is present surrounding the skeletal structures, with the exception of the visceral mass. Cartilage, bone, and other structures stand out clearly.

The author wishes to acknowledge that many of the discussions to be presented here have been obtained from members of the Vertebrate Laboratory at the University of Southern California. One point that everyone seems to agree on is that results of the clearing and staining method may at first be highly variable, and that the causes of such variation are somewhat difficult to understand. Eventually, each individual develops a system that "works" for him. The origin and age of the material may be significant in causing variable results. For example, fresh but unpreserved material, fresh material hardened in alcohol, fresh material hardened in formalin, specimens long preserved in alcohol, and specimens long preserved in formalin may all be successfully treated by one method or another, but probably not by the same method.

THE GENERAL METHOD

This technique follows Davis and Gore (1947) quite closely, and follows Hollister (1934) in part. The method is as follows: (1) If possible, use material originally fixed in 10-percent formalin for about 1 week. However, specimens fixed in 95-percent alcohol for the same amount of time may be used. Older alcoholic specimens are less satisfactory. Remove the viscera and skin from larger specimens, but never remove the skin from small delicate individuals. (2) Transfer specimens to distilled water for 24 hours, in order to soak out the formalin or alcohol. (3) Place specimens in a 2-percent solution of KOH for 12 to 24 hours. (4) Add a 3-percent hydrogen peroxide solution directly to the KOH to remove the pigment. If the specimens are not heavily pigmented, or if the pigment has been removed, go on to the next step. (5) Now, place the specimens in a fresh 2-percent KOH solution, after washing in distilled water for 12 to 24 hours; add 1 cubic centimeter of the stock alizarin red stain to the KOH, immediately. If the hydrogen peroxide treatment was omitted above, the stain may be added to the initial KOH solution.

(6) After 24 to 48 hours, or longer (observe the vertebrae to see how the stain is taking), place the specimens in a fresh solution of 2-percent KOH containing a few drops of glycerin. This will destain and clear the specimens simultaneously. The specimens should be left in this solution until most of the stain is removed and the flesh becomes quite clear. This may take a few days for some individuals and up to $1\frac{1}{2}$ or 2 weeks for others. If the specimens later prove to be unsatisfactorily cleared, they may be returned to the solution for additional clearing.

(7) Finally, the specimens should be transferred to pure glycerin by mov-

ing them through three solutions for 24 hours each, as follows: solution one, 20 cubic centimeters of U.S.P. white glycerin, 3 cubic centimeters of 2-percent KOH and 77 cubic centimeters of water; solution two, 50 cubic centimeters of glycerin, 3 cubic centimeters of 2-percent KOH, and 47 cubic centimeters of water; solution three, 75 cubic centimeters of glycerin and 25 cubic centimeters of water. The specimens are transferred from solution three into pure U.S.P. white glycerin, in which they are stored.

You may have to use the trial-and-error method to start with. One of the trouble points is leaving the specimens in the peroxide solution too long (usually they should be left no more than a couple of hours). Another point which may give trouble is that of not letting the specimens clear and destain long enough.

The stock solution of stain is made according to Hollister (1934) as follows: Make a saturated solution of alizarin in 5 cubic centimeters of glacial acetic acid, 10 cubic centimeters of white glycerin, and 60 cubic centimeters of a 1-percent solution of chloral hydrate.

Some Modifications

Fishes. The above technique works quite well for fishes, with some modifications. Hollister (1934) preferred fresh specimens that had not been preserved or hardened in either formalin or alcohol. Davis and Gore (working with reptiles and amphibians, 1947) found Hollister's method unsatisfactory. The formalin or alcohol method given above is satisfactory here. Fishes should be eviscerated if they are very small. Hollister and others recommend pricking through the skin and scales with a needle to permit the rapid entry of the various solutions and stain, and to prevent the build-up of bubbles under the skin during destaining. Hollister used KOH solutions ranging from 1 to 4 percent. She also found that it was essential to add additional stain for fishes with dense skeletons, since this is absorbed by the calcium salts. If the fish specimens are stout enough, the scales may be scraped off with a knife; if they are very delicate, scaling should wait until the clearing process has been completed. Finally, if difficulty develops in transferring the specimens into pure glycerin, precede the above-mentioned three solutions with the following: solution A, 5 cubic centimeters of glycerin, 90 cubic centimeters of 2-percent KOH and 5 cubic centimeters of water; solution B, 10 cubic centimeters of glycerin, 13 cubic centimeters of 2-percent KOH and 77 cubic centimeters of water.

Amphibians and Reptiles. The above technique works well for amphibians and reptiles. However, one should consult the paper by Davis and Gore (1947) which deals with these animals, for it gives a number of methods worth trying. For example, specimens were hardened in 95-percent alcohol, preferably. Davis and Gore used 4-percent KOH for preserved specimens and exposed these to sunlight for two days prior to bleaching, the latter being done in full-strength hydrogen peroxide. Other workers suggest pin-pricking of reptiles, as with fishes, in preference to skinning, unless ossified material is contained in the skin. Of course, relatively long snakes can readily be skinned and this procedure is recommended.

Birds and Mammals.   Featherless birds and embryonic or newborn mammals may be treated as described above. Otherwise, bird feathers should be removed and mammals should be skinned as described in Chapter 20, with the exception that the feet are left intact. Remove the viscera, eyes, tongue, and other structures that may be too thick to clear well.

STANDARD
SKELETAL TECHNIQUES

Before selecting some method at random, the reader should be advised that the techniques given below for fleshing skeletons do not all produce the same results. Although the objective of these methods is to remove the flesh from the skeleton the degree of efficiency or damage to the skeleton varies. For example, all could be used for preparing disarticulated skeletons, but only one is truly satisfactory for articulated skeletons.

If you are interested in mounting the skeleton on a base or in a glass-topped display box the technique is as follows: (1) Flesh out the specimen and disarticulate it as described below, (2) degrease, (3) bleach, and (4) mount.

Specimens of fish, amphibians, reptiles, birds, and mammals intended for study skeletons should be treated by dermestids if they are too large for the clear and stain technique. Only the largest fish, reptiles and most birds and mammals will require fleshing out. The rest may be treated as described under dermestids below.

Removing Flesh

Fleshing Out.   When dealing with large fish, cut through the skin on the side of the body cavity and remove the viscera. Cut away skin and muscle back along the flanks of the body only in areas where no skeletal structures are involved. Dry and clean, as described under "Dermestid Technique" below. In working with the larger reptiles, birds, and mammals (especially mammals) first identify the specimen and then skin it and remove the viscera. Lucas (1899) recommends carefully detaching the legs from the body, taking special care not to lose the collarbones if these are present. They are small and not articulated in the cat family and very small in the weasel group, but totally lacking in deer, antelope, bears, and seals. Remove the muscle tissue at the point of origin and insertion, taking care not to destroy the ligamentous connections between the bones. Be careful also not to lose the patellae of knee bones of the hind limbs. Next, locate the hyoid apparatus and either remove this immediately or be conscious of it when disarticulating the skull from the spinal column. Disjoint the skull by carefully cutting the ligamentous connections between it and the atlas. Clean out the brain with a brain spoon or a hypodermic, as directed in Chapter 20. Remove the eyes and large bundles of cheek muscles, but take care not to cut soft bony projections from the skull. On larger specimens remove the organs from the rib cage and the flesh from between the ribs, taking care not to cut the cartilage joining the ribs to the breast bone. On very large mammals which must be shipped before cleaning can begin, it is well to separate the vertebral column immediately in front of and behind the rib cage. Unless you are ready to proceed immediately with a cleaning process other than that of the dermestid technique, dry the specimens thoroughly in some place where flies cannot deposit eggs on the flesh. When they are dry, wrap and ship the specimens or proceed with any of the cleaning techniques given below.

Dermestid Technique.   Beetles of the family Dermestidae habitually feed on dead, preferably dry, animal matter, including muscle, hair, feathers, and the like. For this reason they can be extremely destructive if they get into insect collections or bird and mammal collections, but may also be used to clean skeletal material. Dermestids can be obtained from biological supply houses or, more simply, by stopping at the first flat, dry carcass of some bird or mammal that has been killed along the highway. Moist or juicy road kills will attract beetles other than the dermestids, whereas the dry, mummy-like specimens harbor dermestids. Tear open the road kill to locate the specimens. Dermestids may be maintained in a large 5-gallon can with a tight-fitting lid or any other suitable container. This container must provide free ventilation and yet be insect-proof. When not "working," dermestids should be supplied with dog biscuits, dried meat. and other kinds of food material to keep the colony going. It is the larval forms of the dermestids that actually do the cleaning; thus, maintaining a colony means rearing the young of dermestid beetles.

Prepare fish, amphibians, and small reptiles by removing the viscera and completely drying. Keep flies away from the specimens as they dry to keep maggots out of the flesh. Place small vertebrates in with the dermestids, but watch these daily. If the specimens are thoroughly dry the dermestids will first eat the flesh, then will begin to attack the cartilaginous joints between the bones, and finally will begin to attack softer bones, such as those of the fishes. If the flesh is wet inside, however, the dermestids will begin to eat the bones along with the flesh. When the bulk of the flesh has been removed take the skeleton away from the dermestids and finish cleaning with Clorox; this tends to reduce tissue, which can then be picked away with fine forceps.

Less care is required for larger reptiles, birds, and mammals with fully developed bones. When these specimens are thoroughly dried place them in with the dermestids until they are clean. Following cleaning, specimens may be degreased and even bleached if required.

If more than one skull or skeleton at a time is cleaned, each must be fully labeled to prevent confusion. For example, when large numbers of skulls are being cleaned each is placed in a small open cardboard box with its label, and put in with the dermestids. This prevents skulls, jaws, and other parts from becoming mixed up in the container.

Boil-and-Clean Technique.   While many mammologists clean skulls or skeletons by this method, the boil-and-clean technique does more damage to the bones, bone sutures, and ligamentous connections than any other technique. Bone surfaces become very porous, sutures tend to disarticulate, and teeth tend to fall out of their sockets. To use this technique simply boil the specimens until the meat loosens from the bone. Pick the flesh loose, dry the skeleton thoroughly, and then spray the skeleton with a clear plastic spray to reduce the porosity of the bone surface.

Maceration in Water.   If nonligamentous skeletons are needed, simply place the skeletal material, after it has been fleshed out, in a crock of cold water and let this stand at room temperature from 10 to 20 days until the meat is thoroughly rotted and loosened from the bones. Pour off the fluid contents, being careful not to lose any small bones. After drying, degrease and bleach if required. The following technique for preparing ligamentous skeletons for mounting is quoted from *Turtox Service Leaflet No. 9*, 1958, p. 2, by permission of the General Biological Supply House, Incorporated, Chicago.

Place the bones in a glass or earthen-ware container of suitable size and cover the bones with cold tap-water. (Note—never use acids or chemicals of any kind). Do not allow any foreign substance such as metal, wood, etc. to get into the maceration jar as these will discolor the bones. The maceration jar should be kept at room temperature and the water should be changed daily replacing it with fresh water each time. At first a great deal of blood will be evident at each change, but will gradually diminish as it is extracted from the bones. This will take two to three days depending on the temperature of the room. During this time bacterial action (rotting) is taking place on the flesh that was left adhering to the bones. When the bath becomes clear (third or fourth day), pour off and place the bones in a solution made up of one ounce of trisodium phosphate to each gallon of water. Stir well until the trisodium is dissolved, leave the bones in this solution twelve to twenty-four hours. This bath serves a dual purpose in that it halts the maceration and also swells and loosens the tissue remaining on the bones.

Cleaning

Remove the bones from the trisodium phosphate solution and let drain. Secure a small nail or hand brush with stiff bristles, an old toothbrush, chlorinated lime and lots of hot water. Proceed as follows: Dip the brush in hot water, then into the chlorinated lime; now brush the bones using short rapid strokes (use the toothbrush in small hard to get places and on small bones). The action of the hot water and lime plus the friction of the brush creates a burning action upon the tissues adhering to the bones and these virtually disappear as the liming progresses. (Caution—use of rubber gloves in this operation is recommended to protect the skin on the hands from becoming burned.) Rinse in cold water frequently, watching the removal of the flesh carefully. Brush or lime until the flesh has been removed entirely but not the ligaments holding the bones in place. Rinse thoroughly in cold water and lay out to dry at room temperature.

Degreasing Methods

It is not customary to degrease smaller skeletons or skulls unless they are intended for display or student use. Larger bones should have several holes drilled into the shaft and head to permit the degreasing solution to penetrate. Carbon tetrachloride is preferred for degreasing, although white gas works well. As Martin (1964) notes, the solvents are extremely volatile and thus the container should have a small ratio of surface area to volume. He notes that "as both breathing of carbon tetrachloride fumes and absorption through the skin are extremely hazardous, the damaging effects being cumulative even for short, repeated exposures, special care should be taken to insure adequate ventilation through the entire degreasing process and rubber gloves should be worn." A layer of water on top of the carbon tetrachloride will limit evaporation. Specimens should remain in carbon tetrachloride from a few days up to a few weeks, depending on the size of the specimen and the amount of grease. When removing bone specimens carefully tip the container and pour off the grease accumulated on the surface. Martin further recommends covering large skulls with a piece of cloth while they are still under the carbon tetrachloride. When the skull and cloth are removed, only the cloth will pick up surface grease. Air-dry the cleaned bones.

Cleaning and Bleaching

Skeletal specimens that have become dirty from continuous classroom use should be washed with warm water and detergent. Following this, rinse with fresh water and dry. If the bones remain discolored they may be bleached with a 5-percent solution of carbon tetrachloride. If possible, submerge the bones in the solution. Freshly cleaned skeletal material will require up to 10 hours before bleaching is complete. Rinse the bones thoroughly with fresh water and dry.

Display and teaching skeletons are greatly improved if given a light coat of clear plastic spray. This seals the pores, prevents dirt from penetrating, and generally strengthens the bones.

REFERENCES

Davis, D. D., and U. R. Gore, 1947, Clearing and Staining Skeletons of Small Vertebrates, *Fieldiana: Technique No. 4*:3–16.

Evans, H. E., 1948, Clearing and Staining Small Vertebrates, in toto, for Demonstrating Ossification, *Turtox News* 26(2):42–47.

Hollister, G., 1934, Clearing and Dyeing Fish for Bone Study, *Zoologica* 12(10): 89–101.

Knudsen, J. W., 1966, *Biological Techniques: Collecting, Preserving, and Illustrating Plants and Animals,* Harper & Row, New York.

Lucas, F. A., 1899, Notes on the Preparation of Rough Skeletons, *Bull. U.S. Nat. Mus.* 39:1–11.

Martin, R. L., 1964, Skull Degreasing Technique, *Turtox News* 42(10):248–249.

Miller, R. R., 1957, X-Rays as a Tool in Systematic Zoology, *Syst. Zool.* 6(1): 29–40.

Schultze, O., 1897, Uber Herstellung und Conservirung durchsichtiger Embryonen zum Studium der Skeletbildung, *Verhandl. der Anat. Gesellsch: Anat. Anz.* 13:3–5.

Turtox Service Leaflet No. 9, 1958, General Biological Supply House, Inc., Chicago.

Some Display Methods

While the number of biological display methods is practically endless, only a few simple techniques which are applicable to classroom display will be given here.

SPECIMENS PRESERVED
IN LIQUID

Specimens preserved in liquid are displayed in glass jars. Some specimens need not be mounted inside the jar, since they are large enough and firm enough to display themselves. Most specimens, however, are attached to a piece of glass (a glass blank) in order to show the morphology to its greatest advantage. The most desirable display jars are tall, straight-sided museum jars equipped with airtight plastic lids. The 16-ounce and 32-ounce sizes prove most satisfactory. Nothing is more distracting than an entire array of jars of different sizes and shapes on a display shelf. If museum jars are too expensive for classroom display, adopt some common jar which is used generally. For example, many brands of instant coffee come in tall, straight-

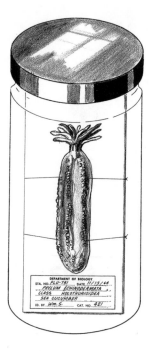

FIG. A–1.
Mounting a specimen in liquid,
on a sheet of glass.

sided, plain jars. If the lids are painted black and a thin layer of soft wax (see Appendix C) is poured on the sealer, these jars prove very satisfactory.

Cut the glass blank (Fig. A–1) as large as possible, to prevent it from moving around in the jar. The blank is roughly ⅘ the inside diameter of the mouth of the jar in width, and is equal to the inside height of the jar. Measure your jar, cut a piece of cardboard to the anticipated size, and test this in the jar to make sure it will lean from the lower front of the jar to the upper back of the jar. Hardware stores with modern glass-cutting tables will cut strips of glass to any desired width. Clear window glass is the most convenient and useful type. Occasionally, opaque black or white glass may enhance the appearance of some specimens. Cut all of the glass blanks to a uniform size so that display bottles will have a uniform appearance.

The specimen label may be glued to the outer surface of the specimen jar or to the lower end of the glass blank. To achieve the latter, roughen the end of the glass blank with sandpaper and glue a standard label to the glass by means of waterproof and alcohol-proof cement, such as Murrayite, Duco Cement, Bond Cement, or others. The label should be printed in India ink, and both the ink and the glue should be permitted to dry for 15 or 20 minutes. Attach the display specimen to the glass blank with black or white thread. Pass a piece of thread through the body of the specimen with a long needle, place the specimen on the glass blank, wrap the thread around the blank (see Fig. A–1) and tie the thread so that it will hold the specimen in the desired position.

Fill the bottle with 70-percent alcohol or 5-percent formalin and lower the glass blank into position. Wipe all preservative away from the mouth and lip of the jar. Finally, put on a lid which has a plastic liner, a soft wax liner, or a piece of polyethylene bag. If the last is to be used, place a piece of polyethylene over the top of the jar, screw on the lid, and cut away the surplus plastic.

DRY SPECIMENS

Modified Insect Cases

Chapter 12 discusses materials and techniques for making insect cases (with or without pinning bottoms). As noted, these can be made to any size, may be equipped with a glass top, and can be made to any depth desired. Many large dried specimens (lobsters, crabs, corals, and so on) are handsomely displayed by attaching these to a piece of heavy, white Bristol drawing board cut to fit inside of the case. Use very fine brass wire to attach the specimen, passing the wire around the specimen and through small holes, perforated in the Bristol board. When the specimen is firmly mounted and labeled, glue the Bristol board into the bottom of the display box.

Insect boxes with pinning bottoms are also excellent for making teaching displays, since labels and specimens may be attached directly to the pinning base. If specimens are properly selected they may be pinned or glued into such a display. Labels may be typed or printed on white paper, and glued to the display. Figure A–2 shows a teaching display constructed by the author.

Riker Mounts

Riker mounts consist of glass-topped boxes filled with pure white cotton upon which specimens and labels are mounted. They are very suitable for insects, plant material, and other medium-sized biological specimens. Chapter 12 gives full instructions for the construction and use of the Riker mount.

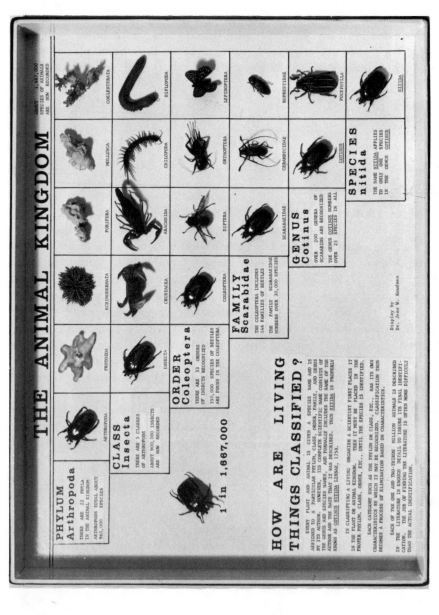

FIG. A-2. A photograph of a display made in a converted insect-mounting box.

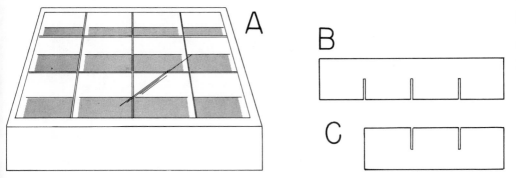

FIG. A–3. Making partitions for glass-topped display boxes.

Partitioned Boxes

Frequently, small to medium-sized specimens are best displayed in a partitioned, glass-topped box such as that shown in Fig. A–3. The box should be constructed like the insect pinning box, as described in Chapter 12. An alternative method is to use the technique for the Riker mount, also given in Chapter 12. For example, if a large supply of uniform boxes is available (such as stationery boxes which are 2 by 9 by 11½ inches), prepare these with a glass top as you would for the Riker mount; spray the outside of the box black and the inside white. Make the partitions out of heavy white cardboard, as shown in Fig. A–3.

Slide-Making

Histological techniques and some of the more elaborate whole-mount techniques are beyond the scope of this book. The main intent has been to present simple slide-making techniques for permanent preservation. There are many excellent texts dealing with histological, histochemical, histopathological and/or microbiological techniques to which you should refer. Some of these are listed in the references at the end of this section.

A DEFINITION
OF TERMS AND TECHNIQUES

Fixation

"Fixation" refers to the preservation of an intact organism, or some portion of that organism, in such a way as to render the cells, tissues, or gross structures most suitable for slide preparations. The more common fixatives, listed in Appendix C, include Bouin's fixative, chrom-acidic fixative, corrosive sublimate, FAA (formal-acetic-alcohol), and Kleinenberg's solution. When fixatives are required, these are discussed in each chapter under methods of killing and preserving specimens.

Staining

Specimens are stained with various dyes in order to make whole structures, certain tissues, certain cells, or certain cellular structures stand out in such a fashion that they may be studied when the specimen is mounted on a microscope slide. Vital stains are those which are mixed with water and can be used in moderate concentrations to stain living organisms or portions of living organisms. Obviously, such stains are very useful in natural history studies. Most stains, however, are used after fixation, while the specimen is held in water or in some solution of alcohol. The more important stains cited in this text are listed in Appendix C.

Dehydration

Dehydration in slide-making refers to the removal of water from specimens to be mounted. This is achieved by replacing the water with alcohol. Dehydration, therefore, generally implies that the specimen be placed in 35-, 50-, 75-, 85-, 95-percent, and absolute alcohol, in order to remove all traces of water. Specimens may be placed in vials or small stacking dishes for dehydration. Small specimens should be left in such vials while the alcohols are subsequently changed. If specimens are large enough, however, they may be lifted from one alcohol solution to the next by means of a wire loop. The amount of time that a specimen must remain in each solution of alcohol depends entirely on the size of the organism and the nature of its tissue. Very delicate specimens should be moved slowly through a long series of alcohol solutions, gradually increasing in strength, to prevent shrinking or contracting the tissues. It is a good practice to use at least two washes of absolute alcohol to ensure that all water is removed. A number of the mounting media listed below require little or no dehydration.

Rehydration

Rehydration simply refers to the return of a specimen to water from some concentration of alcohol. This is the reverse of dehydration and is achieved in the same way. That is to say, the specimen should be moved through increasingly diluted concentrations of alcohol to the final water level.

Clearing

The term "clearing" may be used in more ways than one in connection with slide-making. However, it generally refers to that process which follows complete dehydration and precedes the actual mounting of the specimen. The function of clearing is simply to change opaque or semiopaque material to a glassy-clear nature, so that stained structures become readily visible. Xylene, xylol, oil of cloves, oil of cedar, Terpineol, or even lactophenol may be used with different techniques.

When you add one of the clearing agents to fully dehydrated specimens note carefully whether the remaining alcohol turns slightly milky in color. This signifies that some water still remains in the specimen and in the alcohol, and that the specimen will ultimately turn black and opaque. Should water be detected, return the specimens to fresh absolute alcohol until every trace of water is removed.

Mounting Media

Some of the more common mounting media will be listed here. Others with a limited scope of use are listed separately in various chapters.

Balsam.   Canadian balsam has long been a standard mounting medium. It is a true gum derived from certain trees, and is sold in a dry or in a prepared state. Specimens must be fully dehydrated and cleared before mounting in balsam. Balsam will continue the clearing process after the specimen has been mounted. The refractive index of balsam is quite suitable for most specimens, but such things as diatoms and mites become "lost" in balsam because of the similarity of the refractive indices. Balsam has a tendency to darken with age and, therefore, is sometimes replaced by many synthetic mounting media.

Balsam Substitutes.   Permount, Piccolite, Kleermount, and many other synthetic mounting media are excellent as substitutes for balsam. These media may be thickened or diluted to the desired consistency, dry quickly, and do not have a tendency to darken with age. Like balsam, they are used for specimens which have been completely dehydrated and cleared prior to mounting.

Euparal.   Euparal is an excellent mounting medium for most invertebrate types. It comes both in a clear and a green form (Euparal Vert, used for contrast with hematoxylin stains). Specimens are dehydrated to 90-percent alcohol, transferred to Euparal Essence, and then to Euparal itself, or directly from 90-percent alcohol to Euparal. Euparal clears and mounts simultaneously.

Glycerin Jelly.   Many soft-bodied specimens are mounted in glycerin jelly by first transferring them to glycerin and then placing them in glycerin jelly. The glycerin jelly is heated upon a slide in order to receive the specimen and the coverslip. Glycerin jelly mounts are considered temporary unless the coverslip is sealed with Murrayite or some other waterproof cement.

Turtox CMC Mounting Medium.   There are two forms of the Turtox mounting medium (General Biological Supply House, Chicago), CMC-10 and CMC-S. These media will kill, mount, and clear specimens, and one (CMC-S) will stain specimens after they are mounted. Thus, specimens may be transferred directly from water or alcohol into these mounting media. The author has had good success in adding various dry stains to the Turtox CMC-10 to obtain other colors. For example, fast green, orange G, eosin, methylene blue, and others work quite well.

Techniques for Balsam or Permount.   In many places throughout this book references have been made to Appendix B and the methods of mounting in balsam or Permount. Thus, they will be briefly reviewed herein. Whether or not maceration or staining is required must be determined independently. Dehydrate specimens fully through absolute alcohol, clear, and mount. Use coverslip supports (broken glass slide, etc.) of plastic rings, if necessary.

Slides and Coverslips

Although the standard 1- by 3-inch microscope slide is used for the majority of slide mounts, special microscope slides are available. Slides with concavities may be obtained from any supply house for mounting delicate or thick-bodied specimens. As an alternative (concave slides are expensive), inexpensive plastic rings or pieces of broken coverslip (or slide) may be used to support the coverslip over soft-bodied specimens. Plastic rings, which are made especially for microscope slide use, are first cemented to the slide and the medium and specimen are then added. Still another alternative to using depression slides is that of spinning a cell directly on the slide. For this, a commercial turntable is used. A glass slide is placed on the turntable, and a ring of gold size is spun on the slide to the desired thickness. When this is dry, the medium and specimen are placed in the center of this ring and are covered by a glass coverslip.

You can make your own turntable with a block of wood, a ¼-inch ball bearing, a large spike, the lid of a gallon bottle, and two microscope clips, as shown in Fig. B–1. Remove the sealer from the lid, place a large spike directly in the center of the lid, and pour the lid full of lead. When the lead has hardened, cut the spike off 1½ inches below the lid. Drill a ¼-inch hole in a block of wood, insert a ball bearing, and rest the turntable upon this (Fig. B–1). Finally, drill two small holes to receive the microscope clips. This turntable will work very satisfactorily for making spun cells.

Coverslips or coverglasses come in various sizes, shapes, and thicknesses. Always select glass in preference to plastic coverslips for permanent mounts. The No. 2 thickness (or heavier) is desirable for whole mounts, whereas the No. 1 thickness is essential for thin tissue sections.

Maceration of Tissue—Arthropods

When insects and some other arthropods are mounted for the purpose of studying the exoskeletal structure, it is essential to remove all muscle and other tissues from the specimen. Tissues are macerated either by placing the specimens in a 10-percent solution of KOH for 12 or more hours or by gently boiling the specimen in KOH. In the former technique, you should check specimens periodically under the dissecting microscope to determine when

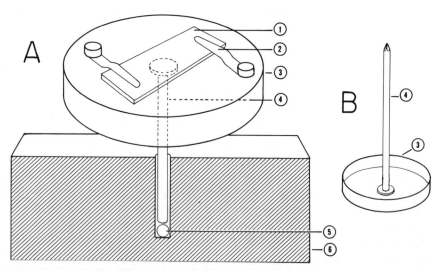

FIG. B–1. A homemade turntable for ringing slides. A. A view of the turntable with the base cut away. B. Forming the turntable. See text. (1) Glass slide, (2) microscope clip, (3) jar lid, (4) large nail, (5) ball bearing, (6) wooden block.

all of the tissues have been removed. In the latter technique, boil the specimens in a small beaker over a hot plate or in a test tube over a Bunsen burner. Use extreme caution, as KOH may be violently expelled from the container. Therefore, always direct the test tube away from you and never hold your face near the mouth of a beaker. When specimens are sufficiently macerated, wash them in several changes of clean water and one change of 30-percent acid alcohol (see Appendix C).

REFERENCES

Baker, J. R., 1958, *Principles of Biological Microtechniques,* Methuen, London.

Conn, H. J., 1953, *Biological Stains,* Biotech Publications, Geneva, N.Y.

Cowdry, E. V., 1952, *Laboratory Technique in Biology and Medicine,* Williams & Wilkins, Baltimore.

Davenport, H. A., 1960, *Histological and Histochemical Techniques,* Saunders, Philadelphia.

Galigher, A. E., 1934, "The Essentials of Practical Microtechnique," published privately.

Gray, P., 1954, *The Microtomist Formulary and Guide,* McGraw-Hill, Blakiston Division, New York.

Gray, P., 1964, *Handbook of Basic Microtechnique,* McGraw-Hill, New York.

Gridley, M. F., 1957, *Manual of Histologic and Special Staining Technics,* Armed Forces Institute of Pathology, Washington, D.C.

Gurr, E., 1956, *A Practical Manual of Medical and Biological Staining Techniques,* Wiley (Interscience), New York.

Humason, G. L., 1967, *Animal Tissue Techniques,* Freeman, San Francisco.

Lillie, R. D., 1954, *Histopathologic Technique and Practical Histochemistry,* McGraw-Hill, Blakiston Division, New York.

Mallory, F. B., 1968, *Pathological Technique,* Hafner, Darien, Conn.

Reagents and Solutions

Only the general and broadly used reagents and solutions are listed herein. Those with limited application are treated in the text: see the Index. For a complete treatment of stains and reagents consult any of the references given in Appendix B.

ALBUMEN (MAYER'S)
> Eggwhite   50 cc.
> Glycerin   50 cc.
> Sodium salicylate   1 gm.
>> Mix together and shake in a clean bottle until emulsified; filter (this may require several days).

ALCOHOL
> The word "alcohol" always refers to ethyl or grain alcohol.

ALCOHOL, ABSOLUTE
> This has *no* water; it is 100% or 200 proof. So-called "100% alcohol" (a trade term) may have up to 0.5% water.

ALCOHOL, ACID
> Alcohol of the proper percentage   100 cc.
> Hydrochloric acid, conc.   6 drops

ALCOHOL, ALKALINE
> Add a few drops of 0.1% sodium bicarbonate to the 70% alcohol wash.

AMMONIA WATER
> Ammonia (NH_4OH)   2 drops
> Water   500 cc.

BERLESE'S MEDIUM (Gray, 1952, see Chapter 13)
> Water   10 cc.
> Acetic acid, glacial   3 cc.
> Dextrose syrup   5 cc.
> Gum acacia   8 gm.
> Chloral hydrate   75 gm.
>
> Mix water, acid, and dextrose. Dissolve the gum in this mixture (requires over a week, with occasional stirring; avoid air bubbles). When in solution, stir in the chloral hydrate.

BOUIN'S FIXATIVE
> Picric acid, saturated aqueous sol.   75 parts
> Formalin, commercial   25 parts
> Acetic acid, glacial   5 parts

CARMINE, ACETO-
> Water   25 cc.
> Acetic acid (glacial)   25 cc.
> Dry carmine stain
>
> Mix acid and water slowly. Add dry stain to mixture in excess of that which initially dissolves, heat to 95° C. for 10 minutes, filter. Use with an equal amount of 70% alcohol.

CARMINE, BORAX- STAIN
> Carmine   1.5 gm.
> Borax   2 gm.
> Water   50 cc.
> Alcohol, 70%   50 cc.
>
> Mix carmine, borax, and water, boil for 30 minutes. Add alcohol, age for 2 days, filter. Very useful for small invertebrates.

CHAMPY'S FIXATIVE
> 3% Potassium dichromate   7 parts
> 1% Chromic acid   7 parts
> 2% Osmium tetroxide   4 parts
>   Mix just before using.

CHROM-ACETIC FIXATIVE
> Chromic acid, 1%   100 cc.
> Acetic acid, glacial   5 cc.

CHROM-OSMIC MIXTURE
>Water   99 cc.
>Chromic acid   1 cc.
>Osmic acid   1 cc. of 1%
>Acetic acid, glacial   10 cc.

CORROSIVE SUBLIMATE
>Saturate water and filter (5 gm. mercuric chloride per 100 cc. water). This is used either hot (50° to 60° C.) or cold. Do not inhale fumes; wash vigorously if it contacts the skin. Do not use metal forceps or containers. For wet specimens, wash after fixing and transfer to alcohol, 50%. Add an iodine solution, drop by drop, until iodine does not lose its color. This will remove remaining corrosive sublimate and prevent specimens from turning black. For dried specimens, do not soak out before drying.

CORROSIVE SUBLIMATE, ACETIC
>Saturated sol. corrosive sublimate   100 cc.
>Acetic acid, glacial   5 cc.

CRYSTAL VIOLET
>Crystal violet   3 gm.
>Distilled water   80 cc.
>Ethyl alcohol   20 cc.
>Ammonium oxalate   0.8 gm.
>   Dissolve crystal violet in alcohol, add ammonium oxalate and water.

DA FANO'S FIXATIVE
>Cobalt nitrate   1 gm.
>Sodium chloride   1 gm.
>Formaldehyde   10 cc.
>Distilled water   90 cc.
>   For field fixation of protozoans, mix at half water with organisms and half fixative; decant after 1 hour and transfer to 70% alcohol.

FAA   FORMAL-ACETIC-ALCOHOL: For plants or animals
>Formaldehyde, commercial   10 parts for animals, 2 parts for plants
>Alcohol, 95%   50 parts
>Acetic acid   2 parts
>Water   40 parts

FORMALIN
>This term always refers to a solution of formaldehyde. Formaldehyde comes in a saturated solution of about 39% or 40%, and thus may be referred to as commercial formaldehyde. *Always* treat commercial formaldehyde as 100% formalin when making a formalin solution. (In other words, 9 parts water and 1 part commercial formaldehyde

make a 10% formalin solution.) Available at drugstores or biological supply houses. Caution, formaldehyde polymerizes when cold; a white powder in the bottle is an indicator. Moderate warmth for storage is advised.

FORMALIN, BUFFERED NEUTRAL

For long-term storage where neutral or slightly basic (pH 7.5) formalin is required, add 6 oz. of hexamine to each quart of formaldehyde.

FORMALIN, NEUTRAL

For general use, add borax or Boraxo, check with litmus paper to be sure that a neutral or basic pH is obtained. Coral may be crushed and added to formalin if nothing else is available.

FUCHSIN, ACID

0.5 gm. dissolved in 100 cc. of distilled water. Acidify with a drop or two of HCl before using. Excellent for Crustacea.

FUCHSIN, CARBOL-

Solution A
    Basic fuchsin   0.3 gm.
    Ethyl alcohol (95%)   10 cc.
Solution B
    Phenol   5 gm.
Distilled water   95 cc.
Mix solutions A and B.

GILSON'S FIXATIVE

Mercuric chloride   5 gm.
Nitric acid, 80% sol.   5 cc.
Acetic acid, glacial   1 cc.
Alcohol, 70%   25 cc.
Water   220 cc.
    Filter after 3 days.

GRAM'S IODINE

Iodine, c.p.   1 gm.
Potassium iodide   2 gm.
Water, distilled   300 cc.
    Grind iodine and potassium iodide in a mortar, add water, transfer to a graduated cylinder, add the remainder of the water, and mix.

GRAY AND WESS' MEDIUM (Gray, 1952, see Chapter 13)

Polyvinyl alcohol   2 gm.
Acetone, 70%   7 cc.
Glycerin   5 cc.
Lactic acid   5 cc.

Water   10 cc.

This has a low refractive index.

HAUG'S SOLUTION (from Gray, 1952)

Alcohol, 95%   70 cc.

Water   30 cc.

Phloroglucinol   1 gm.

Nitric acid   5 cc.

Heat the acid slowly in a warm water bath (never have a flame nearby) and dissolve phloroglucinol into acid. Cool and add to water. Add alcohol and mix.

HEIDENHAIN'S SUSA

Water   90 cc.

Potassium dichromate   1.8 gm.

Mercuric chloride   4.5 gm.

Glacial acetic acid   4.5 cc.

Formaldehyde   10 cc.

For field fixation of protozoans, mix at half water with organisms and half fixative; decant after 1 hour and transfer to 70% alcohol.

HEMATOXYLIN, EHRLICH'S ACID

Hematoxylin   2 gm.

Absolute alcohol   100 cc.

Distilled water   100 cc.

Glycerin   100 cc.

Acetic acid, glacial   25 cc.

Potassium alum   10 gm.

Dissolve hematoxylin in the alcohol and acid. Dissolve alum in heated water. Mix together. Place in stoppered bottle and age until it turns a dark red (up to several weeks). Ready for use. Keeps for years.

HEMATOXYLIN (HARRIS)

Hematoxylin (10% in absolute alcohol)   5 cc.

Potash alum (10% aqueous)   100 cc.

Glacial acetic acid   4 cc.

Mercuric oxide   0.25 gm.

Mix the hematoxylin and alum solutions, heat to boiling point and add the mercuric oxide; when the solution turns deep purple, turn off the heat; cool and add the acetic acid.

HOLLANDE'S FIXATIVE

Picric acid   4 gm.

Neutral copper acetate   2.5 gm.

Formaldehyde   10 cc.

Glacial acetic acid   1.5 cc.

Distilled water   100 cc.

For field fixation of protozoans, mix at half water with organisms and half fixative; decant after 1 hour and transfer to 70% alcohol.

KLEINENBERG'S SOLUTION

Picric acid, saturated sol.   100 cc.

Sulfuric acid, concentrated   2 cc.

Distilled water   300 cc.

Mix acids, filter, add water.

METHOCELLULOSE SLIDE MOUNTING MEDIUM

(after Baker and Wharton, 1952, see Chapter 13)

Methocellulose   5 gm.

Carbowax, 4,000   2 gm.

Diethylene glycol   1 cc.

Alcohol, 95%   25 cc.

Lactic acid   100 cc.

Distilled water   25 cc.

METHYL CELLULOSE

Dissolve 10 gm. of methyl cellulose in 90 cc. of distilled water.

METHYLENE BLUE

Methylene blue   0.3 gm.

Ethyl alcohol   30 cc.

Distilled water   100 cc.

Potassium hydroxide   0.01 gm.

Dissolve stain in alcohol, then mix with water and KOH.

MONK'S MEDIUM

White Karo syrup   5 cc.

Certo (fruit pectin)   5 cc.

Water   3 cc.

Make fresh or add thymol to preserve.

NOLAND'S FIXATIVE

Saturated aqueous solution   80 cc.

Formaldehyde (100% formalin)   20 cc.

Glycerin   4 cc.

Gentian violet (pre-moistened in 1 cc. of water after weighing) 20 mg.

Used with protozoans.

QUINALDINE

Mix 1 part quinaldine with 20 parts of either acetone or isopropyl alcohol. Quinaldine is available from Eastman Chemical Corporation, Rochester, N.Y.

SAFRANIN

Safranin   0.25 gm.

Alcohol, 95% 10 cc.

Water 100 cc.

SALINE FOR COLD-BLOODED VERTEBRATES

Sodium chloride 7 gm.

Distilled water 1000 cc.

SALINE FOR WARM-BLOODED VERTEBRATES

Sodium chloride 8.5 gm.

Distilled water 1000 cc.

SALINE, RINGER'S SOLUTION

Sodium chloride 8 gm.

Sodium bicarbonate 0.2 gm.

Potassium chloride 0.2 gm.

Distilled water 1000 cc.

Calcium chloride 0.2 gm.

Mix calcium chloride with a little of the water. Mix the remainder together. Mix both solutions together.

SCHAUDINN'S FIXATIVE

Alcohol, 95% ethyl, 100 cc.

Mercuric chloride, saturated solution, 200 cc.

Acetic acid, glacial, 15 cc.

SOFT WAX

1 part Paraffin

1 part Vaseline

For sealing coverslips—liquefy with heat and apply to edges of coverslip with a cotton swab.

VITAL STAINS

Methylene blue

Methylene green

Janus green B

Aniline yellow

Neutral red

Crystal violet

Bismarck brown

Mix with water as directed on the container.

WRIGHTS' STAIN

Dry Wrights' stain 0.1 gm.

Methyl alcohol, absolute (acetone-free) 60 cc.

Grind in mortar, mix, filter. Cover blood smear 3–5 minutes; next, add distilled water until a metallic film appears. Let stand 4 minutes, wash in distilled water, and dry.

D

Narcotizing Agents

ALCOHOL (ETHYL). Add, drop by drop, to the culture water over a period of an hour or more until a 5- to 10-percent solution is obtained. Let stand until organisms are insensitive. Otherwise, use as directed.

BENZOCAIN. A trade name for a local anesthetic (actually ethyl amino benzoate) obtainable at drugstores. No permit needed. Add water as directed.

CHLORAL HYDRATE. Used like Chloretone. Obtainable at drugstores or biological supply houses.

CHLORETONE. A Parke-Davis trade name for chlorobutanol obtainable at drugstores or at General Biological Supply House, Chicago. Add water as directed.

CHLOROFORM. Add to culture water by placing tip of pipette under water so that the drops sink. Cover container. Obtainable at drugstores.

CLOVE OIL. Add to culture dish by placing tip of pipette under water, so that drops sink to bottom. Useful for quieting shrimp and narcotizing vertebrates and invertebrates. Obtain at drugstores or biological supply houses.

EPSOM SALTS. This crude form works as well as, or better (echinoderms) than, the pure form, magnesium sulfate. Obtainable at markets or drugstores. Add crystals to water as directed for specific animals.

MENTHOL.   Add crystals to culture water as directed for specific animals; cover container. Obtainable at drugstores and biological supply houses.

MS-222.   This anesthetic (methanesulfonate of *meta*-aminobenzoic acid ethyl ester) is used extensively for quieting and transporting fishes. In excess it will narcotize and even kill specimens in a relaxed condition. Some experimentation has provided excellent results for nudibranchs, other gastropods and some annelids. Obtainable through Canadian and British drug firms generally, and possibly through drug suppliers within the United States.

NICOTINE.   May be useful for many invertebrates. For microscopic organisms, fill a bottle with cigarette smoke and cover opening with an inverted slide containing a drop culture. For large organisms, place tobacco in culture water.

STALE OR PUTRID WATER.   Habitat water previously boiled, or provided with some organic material, and allowed to stand at room temperature, will kill invertebrates, often in an expanded condition, as the oxygen is used up.

INDEX